Custom-Specific Integrated Circuits

ELECTRICAL ENGINEERING AND ELECTRONICS

A Series of Reference Books and Textbooks

Editors

Marlin O. Thurston
Department of Electrical
Engineering
The Ohio State University
Columbus, Ohio

William Middendorf
Department of Electrical
and Computer Engineering
University of Cincinnati
Cincinnati, Ohio

1. Rational Fault Analysis, *edited by Richard Saeks and S. R. Liberty*
2. Nonparametric Methods in Communications, *edited by P. Papantoni-Kazakos and Dimitri Kazakos*
3. Interactive Pattern Recognition, *Yi-tzuu Chien*
4. Solid-State Electronics, *Lawrence E. Murr*
5. Electronic, Magnetic, and Thermal Properties of Solid Materials, *Klaus Schröder*
6. Magnetic-Bubble Memory Technology, *Hsu Chang*
7. Transformer and Inductor Design Handbook, *Colonel Wm. T. McLyman*
8. Electromagnetics: Classical and Modern Theory and Applications, *Samuel Seely and Alexander D. Poularikas*
9. One-Dimensional Digital Signal Processing, *Chi-Tsong Chen*
10. Interconnected Dynamical Systems, *Raymond A. DeCarlo and Richard Saeks*
11. Modern Digital Control Systems, *Raymond G. Jacquot*
12. Hybrid Circuit Design and Manufacture, *Roydn D. Jones*
13. Magnetic Core Selection for Transformers and Inductors: A User's Guide to Practice and Specification, *Colonel Wm. T. McLyman*
14. Static and Rotating Electromagnetic Devices, *Richard H. Engelmann*
15. Energy-Efficient Electric Motors: Selection and Application, *John C. Andreas*
16. Electromagnetic Compossibility, *Heinz M. Schlicke*
17. Electronics: Models, Analysis, and Systems, *James G. Gottling*

Other Volumes in Preparation

Custom-Specific Integrated Circuits

Design and Fabrication

Stanley L. Hurst

School of Electrical Engineering
University of Bath
Bath, England

MARCEL DEKKER, INC.   New York and Basel

Library of Congress Cataloging in Publication Data

Hurst, S. L. (Stanley Leonard)
    Custom-specific integrated circuits.

    (Electrical Engineering and electronics ; 24)
    Includes bibliographies and index.
    1. Integrated circuits--Design and construction.
I. Title. II. Series.
TK7874.H88   1985   621.381'73   84-28719
ISBN 0-8247-7302-0

MARCEL DEKKER, INC.
270 Madison Avenue, New York, New York 10016

Current printing (last digit):
10 9 8 7 6 5 4 3 2 1

PRINTED IN THE UNITED STATES OF AMERICA

Preface

The greatest evolution that has taken place in the short history of present-day technology has been that of the semiconductor industry and the resulting application of microelectronics to an ever-increasing range of activities. It is also remarkable that in spite of increasing sophistication, the cost of electronic equipment is generally decreasing, which is in complete contrast to almost every other commercial and consumer product throughout the world.

However, until recently, the growing expertise of the semiconductor industry in the design and fabrication of microelectronic circuits was not widely available to the industrial original equipment manufacturer or research team, should they be seeking relatively small quantities of a special circuit to match their particular equipment requirements. The expertise of the semiconductor manufacturer was geared largely to the volume production of a range of standard parts, the circuit design of which had been refined to a high degree in order to fully minimize silicon area and thereby reduce production costs and rejection rates, and also to maximize performance to gain market advantages.

The fundamental problem with this mainstream evolution of "hand-crafted" design-on-silicon is that basic design times and design costs were high, thus requiring very large volume production runs in order to reduce the per-unit cost of the final item to an acceptable commercial level. In addition, a mystique surrounding microelectronic design on silicon evolved, largely because of the emphasis on minimizing the silicon area of the final circuit and maximizing circuit performance.

The current situation, however, is rapidly changing. Production techniques are largely becoming standardized, particularly where the

ultimate state-of-the-art performance is not necessary. Further, there
is greater appreciation for the fact that rapid design time is by far the
most critical parameter in reducing overall design and production costs.

In this book we will consider various methods that have evolved in
order to reduce the time to design a microelectronic circuit to fit a particular requirement. All of these methods capitalize on the ease of fabrication of standard-technology microelectronic circuits once the design
steps have been done, but minimize one way or another the amount of
effort required to complete the unique design procedure. While field
programmable devices will be included in our coverage, the greatest
future application will undoubtedly be in masterslice gate arrays and
other custom-specific design methods. Present-day equipment manufacturers are all conversant with the design of printed circuit boards
for their equipment; within the next decade they will undoubtedly and
easily transfer their design expertise to direct-on-silicon circuit and
system design.

We cannot hope to cover all the many varied aspects of the custom-
specific LSI and VLSI area in this one text, or be fully up-to-date with
all ongoing developments. However, I hope that the references provided
here will enable the reader to pursue particular areas of interest in
greater depth and also will give an indication as to the general source
of continuing technical papers in this field.

Stanley L. Hurst

Acknowledgments

The author wishes to acknowledge the financial support during past years of the UK Science and Engineering Research Council (SERC) and the Wolfson Foundation for R and D work undertaken at the University of Bath on masterslice gate arrays and associated software aspects. The work of postgraduate research students working on various aspects of digital logic design theory is also acknowledged. Finally, the author must thank Paul Jennings, formerly Wolfson Research Officer within the School of Electrical Engineering and now on the staff of Imperial College, University of London, for his initiative in many aspects of CAD for semicustom microelectronics.

Contents

List of Symbols and Units

SYMBOLS

Å	angstrom unit, 10^{-10} m, m
C_{ox}	gate oxide capacitance, F
C_O	gate oxide capacitance per unit area, F m^{-2}
E_a	electron energy level of acceptor impurity, eV
E_c	lower electron energy level of the conduction band, eV
E_d	electron energy level of donor impurity, eV
E_F	Fermi electron energy level, eV
E_I	intrinsic electron energy level, eV
E_v	upper electron energy level of the valency band, eV
E_g	electron energy of the valency-to-conduction gap, eV
I_B	dc base current, A
I_C	dc collector current, A
$I_{C(sat)}$	saturated collector current, A
I_E	dc emitter current, A
I_{CBO}	collector-to-base leakage current with $I_E = 0$, A
I_{CEO}	collector-to-emitter leakage current with $I_B = 0$, A

I_D	dc drain current, A
I_G	dc gate current (usually zero), A
I_S	dc source current (usually = I_D), A
$I_{D(sat)}$	saturated drain current, A
I_{DSS}	drain current with V_{GS} = 0, A
K'	MOS process gain factor, AV^{-2}
L	length of conduction channel, m
W	width of conduction channel, m
W/L	aspect ratio of conduction channel
R_{ON}	"on" drain-to-source resistance, Ω
V_{BB}	dc base supply voltage, V
V_{CC}	dc collector supply voltage, V
V_{EE}	dc emitter supply voltage, V
V_{BE}	dc base-to-emitter voltage, V
V_{CB}	dc collector-to-base voltage, V
V_{CE}	dc collector-to-emitter voltage, V
$V_{CE(sat)}$	saturated collector-to-emitter voltage, V
V_{DD}	dc drain supply voltage, V
V_{GG}	dc gate supply voltage, V
V_{SS}	dc source supply voltage, V
V_{DS}	dc drain-to-source voltage, V
V_{GS}	dc gate-to-source voltage, V
$V_{DS(sat)}$	V_{DS} at $I_{D(sat)}$, V
V_{TH}	gate threshold voltage, V
g_m	transconductance (mutual conductance), AV^{-1}
h_{FB}	dc common-base current gain
h_{FE}	dc common-emitter current gain
β	MOS transistor gain factor, AV^{-2}
γ	MOS transistor parameter, $V^{\frac{1}{2}}$
ε_o	absolute permittivity, Fm^{-1}
ε_r	relative permittivity

μ_N drift mobility of electrons, $m^2V^{-1}s^{-1}$

μ_P drift mobility of holes, $m^2V^{-1}s^{-1}$

ψ_B MOS transistor parameter, see Eq. (2.12), V

λ see p. xiv

α Scaling factor for MOS circuits layouts, see Section 2.3.3

UNITS

The geometry size of microelectronics is variously quoted in metric and imperial measure, with units related as follows.

Length

mil, one-thousandth of an inch, = 0.001"
= 25.4 × 10^{-6} m
= 25.4 μm
= 0.0254 mm

micrometer (μm), one millionth of a meter, = 10^{-6} m
= 10^{-4} cm
= 10^{-3} mm
$\simeq$ 0.04 mils

angstrom (Å), 10^{-10}m; 1000 Å = 0.1 μm

Area

1 square mil (mil^2)   = 645 square micrometers
                               (6.45×10^2 μm^2)
1 square micrometer (μm^2) = 0.00155 square mils
                               (1.55×10^{-3} $mils^2$)

Chip perimeters

Chip perimeters may be quoted in "mil-side-average," which is the average of the two outside edge dimensions of the chip., i.e.,

chip size 100 mils × 110 mils = 105 mils-side-average

The corresponding metric equivalent would be mm-side-average, i.e.,

chip size 2.5 mm × 3.0 mm = 2.75 mm-side-average

(Note: Overall chip dimensions are usually quoted in millimeters or mils; device and detailed on-chip geometries are usually quoted in micrometers. The continued industrial use of mils is regrettable.)

λ (lambda)   a dimensional unit specified in the detailed on-chip
              silicon design phase, by which all critical parameters
              are related, e.g., λ may be specified as 2 μm, and
              critical parameters quoted as 2λ, 1.5λ, etc.

Definitions and Abbreviations

The terms custom, full custom, and semicustom may be found in the literature on the microelectronic design with possible varying meaning. Here we use these terms defined as follows:

Custom design
  (or custom-specific
  design, or application-
  specific design

The design of a microelectron circuit to a particular customer's requirement in contrast with off-the-shelf standard items in volume production. Custom design may be divided into the two broad categories following below.

Full custom design

The custom design of a particular customer's requirement, using hand crafting or CAD at device level to achieve the smallest silicon area and highest efficiency design possible.

Semicustom design

The custom design of a particular customer's requirement, using more expeditious means of design than full custom, namely cell library or gate array or some other preprocessed basis.

Note that with the increasing power of CAD workstations and design systems, the divisions between the various techniques are blurring.

Further definitions used in this text are as follows:

Cell library
: A predesigned and fully documented range of commonly used circuits (macros), from which cell-library semicustom design circuits may be assembled.

Dedication
: The design of the dedicated interconnection of a gate array so as to produce a particular custom design.

Gate array
: A regular array of standard circuits ("cells" or "primitives") on chip, which may be fabricated as a standard item up to but not including the final interconnect metallization which completes a particular custom design.

Hand crafting
: The detailed design of an integrated circuit, not using standard cells or macros, but fully designed at transistor level so as to achieve an optimum layout. (Usually reserved for volume production items.)

Macros
: Logic circuit assemblies which perform MSI-level duties, e.g., counters, adders.

Masterslice, masterchip
: A term applied to a gate array or similar standard chip layout, from which a range of custom requirements can be made by the final interconnect design.

Personalization
: The process of committing a programmable device to a particular customer requirement. May be mask or field programmable.

Polycells
: Macros in a cell library (see above), where all the macros are designed to some standard grid dimension, that is, they do not possess random lengths and widths in their silicon layout.

Primitives
: Basic logic circuits from which higher level macros and hence final systems may be assembled.

Programmable read-only memories
Programmable logic arrays
Programmable gate arrays
Programmable logic sequences
Other programmable devices

See Chapter 3, and in particular Table 3.4.

Uncommitted component array (UCA)	A masterslice array where all the standard cells on chip are at component level rather than at a functional level. Chip dedication therefore requires internal-cell metallization to make functional entities, as well as between-cell interconnections.
Uncommitted logic array (ULA)	Usually the same as UCA above (registered trade name of Ferranti Plc., UK).
Uncommitted gate array (UGA)	A masterslice array where all the standard cells on chip are functional entities (primitives). Chip dedication therefore requires only between-cell interconnections.
	(Note: The term "uncommitted" is normally dropped, "gate array" being usually sufficiently explicit.)
Universal logic gate	A functional logic circuit (primitive) which has some defined universal capability and from which all logic requirements may be assembled (e.g., a NAND gate or multiplexer element).

ABBREVIATIONS

CAD	computer-aided design
CAM	computer-aided manufacture
CDI	collector-diffusion isolation
CML	current-mode logic
CMOS	complementary metal-oxide semiconductor
DTL	diode-transistor logic
ECL	emitter-coupled logic
EPROM	erasable programmable read-only memory
EEPROM	electrically erasable programmable read-only memory
EAPROM	electrically alterable programmable read-only memory
FET	field-effect transistor
I^2L	integrated-injection logic
I^3L	isoplanar integrated-injection logic
IGFET	insulated-gate field-effect transistor
ISL	integrated-Schottky logic

JFET	junction field-effect transistor
MOSFET	metal-oxide-silicon field-effect transistor
MTL	merged-transistor logic
nMOS	n-channel metal-oxide-semiconductor
pMOS	p-channel metal-oxide-semiconductor
oem	original equipment manufacturer
PGA	programmable gate array
PLA	programmable logic array
PLS	programmable logic sequencer
PROM	programmable read-only memory
RAM	random-access memory
ROM	read-only memory
RTL	resistor-transistor logic
STL	Schottky-transistor logic
TTL	transistor-transistor logic
UCA	uncommitted component array
UGA	uncommitted gate array
ULA	uncommitted logic array
ULG	universal logic gate
ULM	universal logic module

Custom-Specific Integrated Circuits

1
General Introduction

In this chapter we will attempt to set the scene of present-day microelectronics, to indicate the established strengths of the microelectronic
industry and show the growing evolution whereby other industries and
research organizations may utilize the considerable expertise of the
mainstream microelectronic companies.

1.1 THE ORIGINAL EQUIPMENT MANUFACTURERS' USE
OF MICROELECTRONICS

The use of microelectronic devices by original equipment manufacturers
(OEM) in the design and assembly of their particular equipment lines
spans scarcely 25 years. In the late 1950s and early 1960s, commercial
equipment, often of considerable complexity, was usually designed
using discrete devices. This resulted in considerable emphasis being
placed on peripheral printed circuit board (PCB) and connector developments. Figure 1.1 shows an example of discrete-component solid-state
equipment of the late 1950s. Early semiconductor devices used for production equipment were germanium, which were superceded by silicon
devices around 1961. The transition from germanium to silicon technology was not entirely painless to the original equipment manufacturer.
Germanium transistors had been developed to give higher large-signal
current gains than early silicon transistors, and reverse-bias V_{BE}
breakdown voltages were superior to silicon. Indeed they were, and
in certain respects still are, better solid-state switches than their silicon counterparts.

1

FIGURE 1.1 An early solid-state equipment of 1959/1960 using discrete
devices, for the remote control and supervision of railroad traffic. (Photo
courtesy of Westinghouse Brake and Signal Co., Chippenham, England.)

However, the fundamental temperature limitations posed by germanium were always a serious problem, and silicon technology, particularly planar technology, soon predominated. Rapid evolution of
silicon technology, from initial discrete devices to small-scale integrated (SSI) circuits followed. By mid-1965 discrete-component commercial equipment was the exception rather than the rule, except for
particular applications involving high frequencies or power.*

The increasing expertise of the semiconductor manufacturer with
decreasing device geometries and improvement in device performance

---

*One very significant exception to this statement was of course the
radio and television industry; indeed discrete devices still remain
prominent in this area.

enabled the bipolar evolution through resistor-transistor-logic (RTL), diode-transistor-logic (DTL), and transistor-transistor-logic (TTL) to be pursued, the technical emphasis being upon increased voltage and current-gain availability per circuit, resulting in improved overall system performance and stability. Table 1.1 gives a broad picture of this

TABLE 1.1   The Evolution of TTL 7400-Series Logic Families as Standard Off-The-Shelf Components

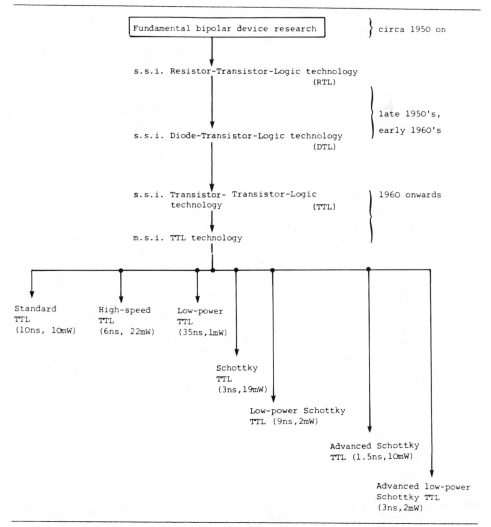

[a]s.s.i = up to 100 transistors per package.   m.s.i. = 100 to 1000 transistors per package.

TTL evolution, which still continues, with second- and third-generation TTL families incorporating advanced low-power Schottky technology to improve performance.

International acceptance of bipolar TTL as an off-the-shelf family of packages was due to availability rather than any overriding intrinsic merit. Fundamental disadvantages include transistor saturation, which implies higher power losses and lower switching speeds in comparison with a nonsaturating technology, and the specification of 5 V as the supply voltage. The later would not be a value any system designer would prefer. However, 5 V has become a standard from which it is now difficult to move, except possibly in the semicustom area which is the main theme considered in this text.

Leaving aside for the moment the more specialized bipolar developments such as nonsaturating emitter-coupled-logic (ECL), the main parallel to these TTL developments as far as the original equipment manufacturer is concerned has been the unipolar metal-oxide semiconductor (MOS) and complementary MOS (CMOS) developments.

The most significant characteristic of MOS compared with bipolar technology is the higher impedance levels of the circuits, which results in lower power dissipation on-chip for the same duty. Hence, more circuitry can be fabricated on-chip without reaching heat dissipation limits, and overall power requirements and dissipation problems per equipment are reduced. The use of complementary MOS reduces the standby dissipation per logic circuit to (ideally) zero, which is a prerequisite for many portable types of equipment. Against this must be weighed the generally slower switching speeds in comparison with the fastest available in alternative technologies, due to the higher impedances which have to drive circuit capacitances between the logic voltage levels.

In spite of the basic simplicity of MOS structures, the commercial evolution of off-the-shelf MOS circuits has not been as rapid as the complex bipolar technology. Indeed, MOS and complementary MOS developments have been somewhat erratic, but possibly have been the spur to maintain the alternative continuing evolution of TTL. Table 1.2 indicates the general progress of MOS developments, from which has largely sprung the current application to semicustom microelectronics.

Tables 1.1 and 1.2, therefore, illustrate the principal avenues of semiconductor technology that have provided off-the-shelf circuits for original equipment manufacturers' use and research and development activities [1-22]. The bipolar area has several families for more specific applicatons in addition to the classic 7400 hierarchy summarized in Table 1.1, including:

1. Current-mode logic (CML), a medium-speed bipolar process originated by Bell Labs, and adopted particulary by Ferranti for semicustom applications with analog and digital capability [23-25].

TABLE 1.2 Avenues of MOS Technology Development[a]

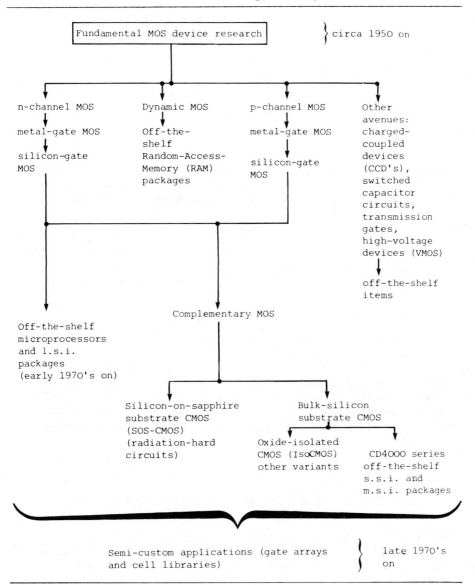

[a]Not necessarily shown in exact relative order of achievement.

TABLE 1.3   Typical Features of Standard Products[a]

	TTL	nMOS	CMOS
Masks per mask set	8	6	7/8
Diffusion and implants	5	3	5
Silicon area per gate (μm^2)	3×10^3	1×10^3	2×10^3
Propagation delay per gate (ns)	5	30	Metal-gate 20 silicon-gate 5
Static power per gate (mW)	10	0.2	>0.0001
Speed-power product (pJ)	50	3	0.5/2

[a]Note: higher-speed and/or smaller geometry versions may be available.

2. Integrated-injection-logic (I^2L), a further medium-speed bipolar process with high packing densities, used by several companies for semicustom applications, again with analog and digital capability [17, 21, 26-29].
3. Emitter-coupled-logic (ECL),* a nonsaturating small-voltage swing, high-power dissipation bipolar technology, for very high speed applications, such as mainframe computers and associated semicustom peripherals [17, 21, 28, 30, 31].

We will refer to these three technologies in subsequent chapters, although they will not be in the mainstream of our interests. In general, the standard product lines provide the equipment manufacturer with off-the-shelf circuits with the typical manufacturing and performance features given in Table 1.3. For further details, see Muroga [28] or the many detailed references contained therein.

1.2   THE USE OF STANDARD OFF-THE-SHELF PARTS

It is normally economically advantageous for an original equipment manufacturer to use standard parts wherever possible [8]. Off-the-shelf microelectronic packages offer the following advantages:

---

*ECL is also a current-mode form of logic. However CML is usually used to refer to technology (1) above as used by Ferranti Ltd. and Interdesign, Inc., while ECL is used to refer to the very high-speed, nonsaturated, emitter-coupled technologies of other companies.

1. Low cost due to volume sales absorbing the initial silicon design costs
2. Optimized on-chip performance, since once-only design time and costs can be afforded to produce an optimal silicon layout
3. High reliability, again due largely to volume production runs
4. Multiple-source availability, so that the equipment manufacturer is not dependent upon one source of supply

However, against these powerful advantages must be weighed certain disadvantages:

1. There is little commercial security in the final equipment, since copying can be readily undertaken
2. The individual off-the-shelf packages may contain more features and input/output (I/O) pins than are required for a specific application
3. No available package provides exactly the requirements of the system being designed, and, therefore, an escalation in the number of standard packages becomes necessary

Hence, in total, a considerably greater cumulative silicon area and input/output package connections may accrue using standard packages than if the precise system requirements were available, and although cost per package may be low, total system design and manufacturing cost may continue to be high. We will consider such economic factors in Sec. 1.7.

The dichotomy between attempting to market a standard volume-production part, with the particular technical advantages itemized above, and making such a standard product have the widest possible application to a range of potential users, was the problem facing the integrated circuit industry in the early 1970s, when large-scale integration (LSI) had become established. At small-and medium-scale integration levels, a considerable range of standard circuits had been proposed, as witnessed by Table 1.4, but the more powerful the integrated circuit the less likely it was to satisfy a wide range of potential users. Thus, a range of totally dedicated packages which utilized the full LSI expertise and capabilities of the semiconductor industry was difficult to propose.

The industry solution of the 1970s was the introduction of the microprocessor, which provided a powerful computational package at low cost, ideal for the integrated circuit manufacturer (vendor), the dedication of the circuit being passed over to the equipment manufacturer (customer) to undertake in the form of software programming. To a large extent the costs of equipment hardware design were replaced by corresponding if not greater costs of software preparation, and it is debatable whether all the tasks to which microprocessors have been

TABLE 1.4 Typical Off-the-Shelf SSI/MSI Circuits Available in TTL and MOS Series, Frequently Necessary as Logic "Glue" Around Larger LSI Packages

Standard function	TTL	CMOS
2 Input NOR	7402	4001
3 Input NOR	7427	4025
4 Input NOR	7423	4002
2 Input NAND	7400	4011
3 Input NAND	7410	4023
4 Input NAND	7420	4012
2 Input OR	7432	4071
3 Input OR	—	4075
4 Input OR	—	4072
2 Input AND	7408	4081
3 Input AND	7411	4073
4 Input AND	7421	4082
2 Input Exclusive-OR	7486	4070
2 Input Exclusive-NOR	7435	4077
AND-NOR, 2 Wide-2 Input	7450	4085
AND-NOR, 2 Wide-3 Input	7451	—
AND-NOR, 4 Wide-2 Input	7453	—
AND-NOR, 2 Wide-4 Input	7455	4086
D Type with PR and CLR	7474	4013
D Type with CLR	74175	4175
JK with PR and CLR	7476	4027
JK with CLR	7473	—
8-Bit addressable latch	74259	4099
4-Bit parallel I/O shift register	74178	40194
8-Bit parallel I/O shift register	74165	4034
8-Bit serial I/O shift register	7491	—
8-Bit serial in/parallel or serial out shift register	74164	40164
4-Bit ripple binary counter	7493	—
4-Bit sync. binary counter	74163	40163
4-Bit sync. UP/DOWN binary counter	74193	4516
4-Bit sync. BCD counter	74162	40162

TABLE 1.4   (Continued)

Standard function	TTL	CMOS
4-Bit sync. UP/DOWN BCD counter	74190	4510
Decade counter	7490	–
Divide-by-12 counter	7492	–
Presettable divide-by-N counter	74163	4018
2-to-1 Multiplexer	74157	
4-to-1 Multiplexer	74153	
1-of-8 Decoder/demultiplexer	74138	
1-of-4 Decoder/demultiplexer	74139	
8 Input multiplexer	74151	
8-Bit priority encoder	74148	

applied have been cost effective in comparison with totally dedicated hardware alternatives. The main points against microprocessor-based equipments are:

1. Software design costs are high, particularly when capital costs of emulators and other design aids are amortized
2. Software reliability may be suspect under unforeseen operating conditions
3. Detailed documentation for maintenance and end-user is expensive
4. Turnover of programming manpower causes problems of continuity and update
5. Commercial security is poor, since software residing in read-only memory can be readily copied
6. Performance is not as high as with dedicated hardware, since part of the microprocessor computing time and power is necessary for its own data organization.

Against these drawbacks must be placed the advantages, the overriding one being flexibility, enabling the same hardware equipment to be used for several variations of duty by appropriate revision of its software programming only.

Thus, where appropriate, the original equipment manufacturer will continue to employ standard off-the-shelf parts produced by the microelectronics industry, using the benefits of the developments previously summarized. Equipment lines such as those illustrated in Figure 1.2 based upon standard products will continue to be commonly available. However, the OEM will also be increasingly concerned with the merits of custom and semicustom techniques, which will be our main theme in the subsequent sections.

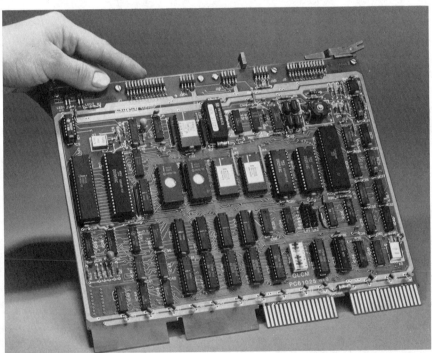

FIGURE 1.2 A present day measurement and control system, using off-the-shelf microelectronic packages. (Photos courtesy of Oxford Automation Ltd., Milton Keynes, England.)

1.3 FULL CUSTOM DESIGN

Having briefly reviewed the advantages and disadvantages of standard off-the-shelf integrated circuit packages, the alternatives which are available to the original equipment manufacturer may now be considered.

We should first define "custom," "full custom," and "semicustom," as these terms will be used in this text. Unfortunately, there is still some lack of common agreement regarding the use of these terms, but here we will use the meanings as defined in the *Definitions and Abbreviations* preface section. The generic term "custom design" will in general be avoided, with the more specific terms full custom and semi-custom design used where appropriate.

Table 1.5 defines the broad divisions of microelectronic design and manufacture. We will consider programmable devices and the semicustom design area later, but first let us discuss the implication of the full custom design route.

In a full custom design approach, the objective is to produce the most efficient final design-on-silicon possible within given time or cost constraints. As will be seen later, the semicustom design routes do not generally offer the most efficient use of silicon (smallest possible chip area) for a given requirement, or the highest available speed performance, and for this reason the equipment manufacturer may wish to consider full custom possibilities, particulary for avionics, space, or military requirements, where cost is not an overriding parameter.

Full custom design entails the circuit and layout design by persons expert in the detailed design of gate structures and interconnect topology at the silicon level. Among the factors involved are:

1. Incorporating only circuitry which is functionally necessary, with no unnecessary or unused gates, interconnections or input/output buffers
2. Possible use of different device geometry sizes in different parts of the circuit in order to maximize performance and/or minimize silicon area
3. Interactive CAD/designer placement-and-routing of the chip in order to minimize silicon area and any critical on-chip interconnect lengths
4. The possible introduction of novel gate structures to provide specific combinatorial logic requirements.

It will be apparent that this procedure is within the expertise of the microelectronic companies, since it is identical to the design procedure for their standard volume products (Figure 1.3). However, it will be beyond the capabilities of an original equipment manufacturer without ongoing microelectronic design experience.

TABLE 1.5   Broad Avenues of Microelectronic Circuit Design and Manufacture

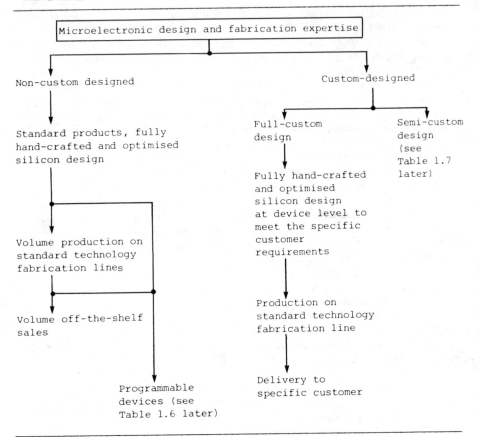

    Hence we have a situation similar to that of many industries should a customer require an entirely handbuilt product, be it a car, a piece of furniture, or an integrated circuit. The cost will be high, but, hopefully, it will meet the customer's requirements.

    Looking more closely at costs, it will be apparent that with the full custom circuit the cost will depend heavily upon the design time and manpower involved; the final fabrication costs will not be excessive as fabrication will (in general) still be done on standard volume-production lines when spare capacity is available. Thus, it is the silicon design costs which will always be the insuperable problem to full custom design in comparison with semicustom design routes; the conflict between

High design costs,   Lower semicustom design
most efficient final   vs. costs, not so efficient final
circuit performance   circuit performance

will always be present.

FIGURE 1.3   A fully handcrafted volume-production chip, involving
many man-years of detailed design and floorplan layout. (Photo courtesy of Intel Corporation, Santa Clara, CA.)

   We will consider costs in greater detail in Sec. 1.7.   Suffice to
state here that full custom design costs for specific requirements
can only be considered when certain overriding considerations apply,
such as for avionic or military requirements.   As a result we will have
few occasions to refer to full custom in this text.
   Details of silicon design techniques are outside our present interests,
but may be found in many publications [4, 14, 28, 32].

1.4 PROGRAMMABLE DEVICES

Programmable devices may be regarded as a half-way stage between a
standard volume-production product, and a custom or semicustom special. All such devices are standard production products up to the final
stage of commitment to a particular custom requirement. This final
commitment is usually arranged either by making certain final connections on the chip (mask-programmable), or by destroying connections
already made on the chip (field-programmable).
   Table 1.6 gives the general hierarchy of programmable devices.
Mask-programmable implies that the final commitment stage for a

TABLE 1.6 Hierarchy of Mask-Programmable and Field-Programmable Microelectronic Packages

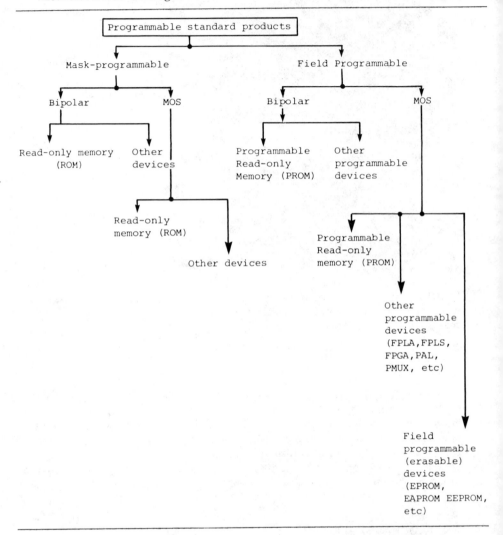

Note: Particular suppliers may have yet further names and abbreviations for their products.

particular requirement is performed by the device manufacturer to meet the customer's specification before shipping, while *field-programmable* implies that the customer is able to commit the device to his requirements, the manufacturer shipping a standard uncommitted off-the-shelf product.

Here we define the principal categories of programmable devices which are currently on the market. In Chapter 3 we will consider each of the principal field-programmable devices in greater detail, but we will not consider mask-programmable devices beyond this section in any great depth. Definitions are expanded below.

1.4.1 Read-Only Memories

Read-only memories (ROMs) are combinatorial logic arrays, in which a number of output functions may be provided for a given set of binary input variables. The term "memory" applied to such circuits is possibly unfortunate, in that no storage in the accepted sense of bistable circuits or other form of latch is incorporated, the outputs being entirely dependent upon the present state of the input variables. In general, the output functions provided by an ROM are fixed at the commitment stage, and do not alter during the lifetime of the circuit. The exception to this is the reprogrammable (erasable) type of ROM, in which the internal connections may be reprogrammed by appropriate means.

The distinguishing feature of ROMs is the regular matrix organization of their internal circuitry, whereby output functions may be synthesized by appropriate connections between orthogonal signal paths. These internal connections may either be added to an unconnected circuit (mask-programmable), or unwanted connections destroyed on a fully connected circuit (field-programmable). A further feature of ROMs is that they are nonvolatile devices, that is, they do not lose their input/output dedication in the event of removal of the direct-current (dc) power supply to them. This is in contrast to random-access memories (RAMs), which are true memory devices containing some form of latch or other electrical storage circuit, with read-in and readout facilities, which lose their existing memory content in the event of supply failure (volatile memory).

Mask-programmable ROMs are generally referred to simply as ROMs. Field-programmable types, however, are normally termed PROMs. Within the latter category we find further definitions such as: EPROM, erasable-programmable ROM; EAROM, electrically alterable ROM; EEPROM, electrically erasable ROM. We will consider the structure and general application of ROMs in Chapter 3. Additional information may be found in many published sources [3, 14, 21, 28, 32-36].

1.4.2 Programmable Logic Arrays

Programmable logic arrays (PLAs) are also combinatorial logic arrays, whereby output functions dependent upon a set of binary input variables may be synthesized. Like ROMs, they are nonvolatile circuit elements, which may be mask-programmed by the supplier, or field-programmed by the user. The former is usually referred to as merely PLA, while the latter is usually referred to as field-programmable logic array (FPLA).

Like ROMs, PLAs are characterized by a regular matrix organization of their internal connections. Unlike ROMs, which, as we will see in Chapter 3, generate output functions in a truth table sum-of-minterms form, PLAs generate their outputs in an appropriate sum-of-products form, where the product terms do not necessarily contain all the input variables. As a result, the use of PLAs normally requires some minimization of the required output functions from a minterm truth table statement of each function, but the effect of such minimization is that more binary input variables are normally available on a PLA structure compared with that of an ROM structure.

However, while a ROM dedication may be designed and verified on a RAM structure, the fixed ROM dedication being a functional copy of the proven RAM read/write design, the dissimilar internal structure of the PLA compared with the minterm-based ROMs and RAMs means that there is no copy facility available between RAM and PLA dedications. PLAs are therefore not usually considered as "memory" devices in the same sense as ROMs, but are purely LSI combinatorial logic packages for random logic applications.

For further details see Chapter 3 or the many published references [14, 28, 32-42]. Reprogrammable PLAs have been reported [43], but are not commonly available.

1.4.3 Further Programmable Circuits

A number of further circuits are available on the commercial market, many with designations specific to the particular integrated circuit (IC) manufacturer. The majority are variants on PLAs, with one distinctive addition in certain cases, namely, the inclusion of clocked bistable storage elements on-chip, the set/reset of these elements being controlled by the combinatorial dedication of the chip, plus provision of a clock input. Hence, the design of sequential networks, counters, random sequences, as well as combinatorial logic requirements may be undertaken.

The designations which may be encountered include the following:

Field-programmable gate array (FPGA). A simplified form of PLA, which has the capability of realizing product (AND and NAND) terms from its binary input variables, but without the direct capability to sum (OR) the product terms together.

Field-programmable logic sequencer (FPLS). An extended form of PLA, with internal clocked storage elements, usually type D or JK circuits, in addition to the sum-of-products combinatorial capability. The D or JK clock-steering inputs of these circuit elements are programmable by the on-chip product array.

Programmable array logic (PAL). Similar to the FPGA,* except that some versions contain storage capability as well as the combinatorial product term capability. Designation specific to certain IC manufacturers (see Chap. 3).

Hard-wired array logic (HAL). Mask-programmable versions of field-programmable PAL devices.*

Programmable multiplexer logic (PMUX). A field-programmable circuit for random combinatorial logic requirements, organized on multiplexer principles to provide each programmable output. It may be useful as a more direct replacement of SSI/MSI multiplexer assemblies than a PROM or PLA equivalent.

All of the above may be considered variants on the PLA concept, tailored to more specific applications than the general purpose PLA sum-of-product array. At present, the majority are field-programmable rather than mask-programmable, with reprogrammable versions not yet announced. For further details, see published references [39-44] and Chapter 3.

1.5 SEMICUSTOM TECHNIQUES

Section 1.3 and Table 1.5 differentiated between standard products, and the full custom and semicustom routes for the supply of integrated circuits specific to a particular customer. Full custom was discussed, with the conclusion that semicustom techniques provided the only generally viable route for the design and manufacture of integrated circuits specific to a particular customer requirement. Here we will begin to consider semicustom design techniques, defining the main principles and subject areas, before returning to the subject in greater depth in Chapters 4 and 5.

Continuing from Table 1.5, the main semicustom design routes are indicated in Table 1.7, namely (1) the cell-library approach, and (2) the uncommitted masterslice (or "masterchip") approach. The principal distinguishing feature of these two alternatives is that a full mask set has to be generated for the production of cell-library designed circuits, but only the final interconnection masks have to be generated for masterslice circuits.

Considering the cell-library approach further, a prerequisite is that there shall already be available a set of proven standard circuits designs,

---

*PAL and HAL are registered trademarks of Monolithic Memories, Inc.

TABLE 1.7   The Two Main Semicustom Design Avenues Whereby
an Original-Equipment Manufacturer May Obtain Circuits to Specific
Requirements

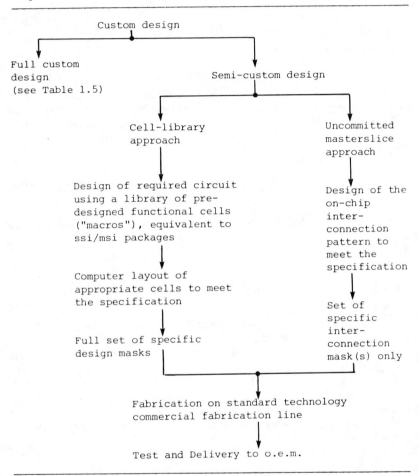

such as gates, latches, and other functional entities from which the
specific custom requirement may be assembled. It may be appreciated
that these available designs, the "cells" or "macros," correspond closely
to off-the-shelf small-scale integrated (SSI) or medium-scale integrated
(MSI) packages which the customer may himself use should he be designing his requirements as a printed circuit board assembly; here,
however, we are not physically using separate circuits, but instead
are connecting together appropriate cell designs to form the complete
monolithic integrated circuit. Figure 1.4 shows a typical cell-library
approach for semicustom applications.

The number of cell designs in the designer's library may vary from a few basic macros to a large number, depending upon resources. Each new customer's design may justify adding further macros to the existing library, and hence a body of macro availability is accumulated. It should be noted that each macro must be fully designed through all the mask levels necessary for its final fabrication, and its performance must be fully documented so that overall system performance may be checked as far as possible at the design stage.

Hence, the cell-library semicustom design procedure will generally be as follows:

1. Partition the customer's requirements into functional macros corresponding to available designs in the cell library
2. Lay out the required cells and design their required interconnections, including final input and output connections (placement and routing)
3. Simulate the overall performance of the complete circuit to check for logic errors, race hazards, etc.
4. Obtain customer's assurance that the design meets his requirements
5. Generate the required complete set of masks to fabricate the unique assembly of macros plus their interconnect, and fabricate in the conventional manner

The financial saving in the semicustom cell-library approach compared with full custom is in the elimination of the very costly manpower involvement in detailed silicon design at transistor level, when every transistor and every part of the circuit is individually handcrafted to achieve maximum circuit efficiency and minimum silicon area. The final silicon area of a cell-library design, however, may be 30—50% greater than a corresponding full custom design, particularly if the available macros do not precisely match the optimum partitioning of the customer's requirements, or if the placement and routing procedure is not optimized due to time or other constraints.

A problem with the cell-library approach is the updating of all designs in the library, which may become desirable or necessary because of changes in fabrication technology. Equally, if the semicustom designer wishes to offer cell-library designs in more than one technology, for example, in single-channel MOS and in CMOS, then he must prepare the fully detailed designs for his cells in both technologies. Thus, a considerable housekeeping exercise may be involved behind the scenes, the cost of which must ultimately be reflected in the final price of the custom-designed product.

We will consider commercial economics further in Sec. 1.7, and Chapter 5 will consider in greater depth the technical features of this whole area. Further general details may be found in published literature [14, 25, 28, 32, 45-47].

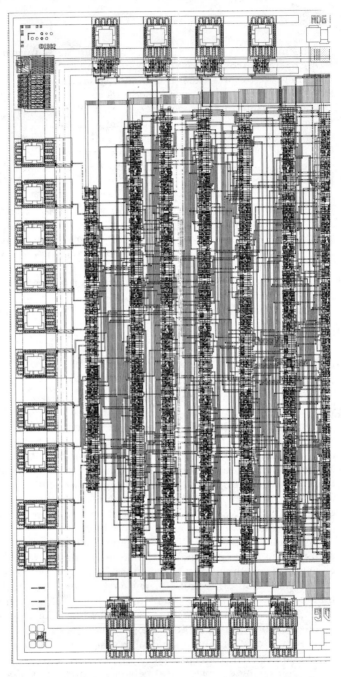

FIGURE 1.4 A commercial cell-library nMOS semicustom LSI circuit, approximately 500 primitive-based cells. (Photo courtesy of Tektronix Inc., Beaverton, OR.)

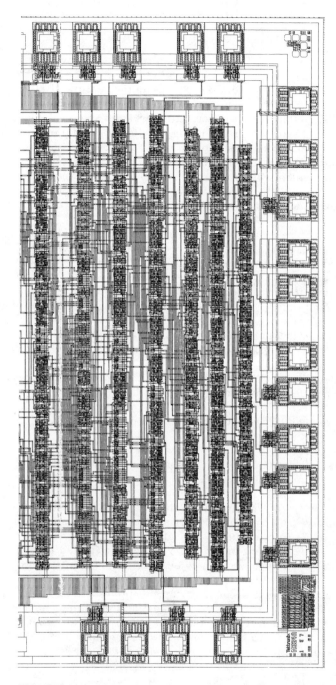

FIGURE 1.4 (Continued)

21

The alternative "masterslice" semicustom approach has many possible technical variations. The terminology associated with this area is also somewhat diverse and nonstandardized. The essential feature of all masterslice (or "masterchip") semicustom realizations in the use of a standard LSI circuit containing an uncommitted array of identical cells, which may be fabricated up to but excluding the final metalization interconnect as a basis for all custom applications. The custom dedication of such a standard array lies in the design of the final interconnection metalization, which will be unique for each customer requirement (Table 1.7). Hence, the uncommitted masterslice wafers may be fabricated in quantity and held in stock to await final specific dedication.

The diversity of commercial masterslice products, and the associated terminology, arises from the number of choices of circuit which may be used as the standard cell in a masterslice array. Two initial choices are available, (1) whether to provide cells containing uncommitted components such as separate transistors, or (2) to provide cells which are each a fully functional entity, such as a NAND or NOR gate. Table 1.8 illustrates this divergence and its implications.

Arising from these variations we have certain terminologies defined as follows:

Uncommitted logic array (ULA)*: An uncommitted masterslice array, with cells at component level rather than functional level.

Uncommitted component array (UCA): A more precise term for a ULA.

Uncommitted gate array (UGA): An uncommitted masterslice array, with cells at function gate level.

The term "gate array" or "masterslice gate array" without further qualification is loosely used to cover all forms of masterslice, and more widely to refer to UGAs in particular. We will attempt to use the terminology uncommitted component array (UCA) and uncommitted gate array (UGA) in the text where we wish to be specific between the two main categories.

From these preliminary discussions, it may be apparent that the masterslice approach should provide the quickest and most inexpensive route to the design and fabrication of a custom LSI circuit, since all the levels of design and fabrication except the final on-chip interconnection are standard. This is generally true, although decreasing

---

*ULA is a registered trademark of Ferranti plc., UK, for semiconductor devices.

TABLE 1.8 Variations in the Masterslice Semicustom Approach

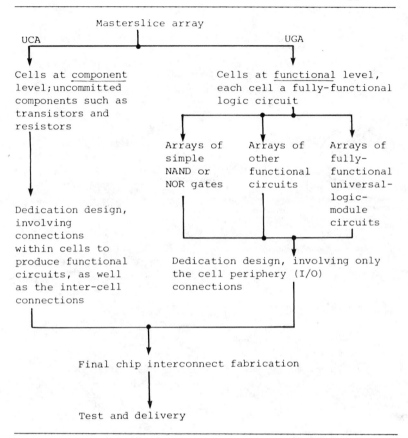

CAD costs and commercial interests may blur differences between masterslice and cell-library design costs. However, the masterslice approach cannot provide such an efficient final circuit as the cell-library approach, when (1) minimization of chip size and (2) optimization of performance are considered. This is because one has to use a particular array size on a masterslice, which must be large enough to contain the custom requirements, but which is likely to contain unused cells in the final design. This is in contrast to the cell-library approach, where only the required circuits (macros) are assembled on the chip, and spacing between the circuits is only that required for the specific interconnections rather than predetermined fixed-width wiring channels as on uncommitted arrays. Figure 1.5 shows a typical uncommitted masterslice gate array, from which the wiring channel areas between rows of cells may be seen.

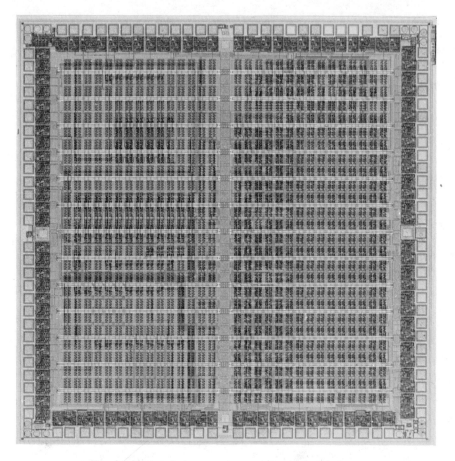

FIGURE 1.5   A commercial masterslice uncommitted gate-array. (Photo
courtesy of Harris Corp., Melbourne, FL.)

Further details of masterslice arrays have been published [14, 25,
28, 32, 45, 48-57]. We may summarize this general introductory section on semicustom techniques before going into further technicalities
as follows:

Cell-library techniques. Dependent upon a predesigned library of
functional circuits, the on-chip placement and routing of which provides a semicustom design of optimum size and performance. Possibly up to 50% larger chip size than a handcrafted full custom circuit for the same specification, but design time considerably less
than full custom.

Masterslice techniques. Dependent upon predesigned component-level or gate-level uncommitted chips, the routing of which provides a semicustom design with the minimum of time and expenditure. Possibly up to twice the silicon area of a handcrafted full custom circuit for the same specification; exact area relationships depend upon the percentage cell utilization on the uncommitted chip.

As a final general comment, we should make the important point that the design-on-silicon, be it full custom, semicustom cell-library,

TABLE 1.9   Flow Chart of Typical Design Procedure for Custom or Semicustom Product

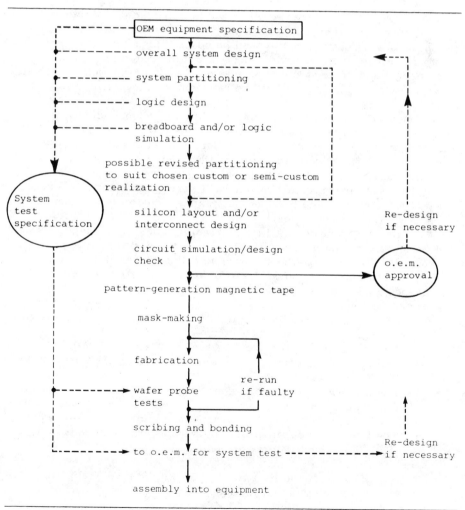

or semicustom masterslice, is only a part of the complete engineering
design from system requirements to the final packaged product. The
total number of steps in the system design procedure may be as shown
in Table 1.9; the different design-on-silicon techniques all fit into this
general structure, each having problems of ensuring that the different
groups of engineers down the chain of design are correctly interpreting
the original system requirements. We shall have more to say about
interface problems and communication of requirements.

1.6 CONSTRAINTS AND LIMITATIONS

Apart from economic factors, which will be considered separately (Sec.
1.7), there are other constraints and limitations which dictate whether
custom or semicustom circuits shall be adopted for a particular application. All these factors and aspects are interrelated and continually
changing, and no hard and fast rules can be given when off-the-shelf
microelectronic products should be retained, or alternative avenues
exploited.

If new equipment has to be produced by an original equipment manufacturer within a very short time, he may be reluctant to subcontract
a vital part of his microelectronics to an outside company upon whom
he cannot exert the same pressure as within his own organization.
Thus, the time factor involved in obtaining custom or semicustom parts
may be a serious concern. Certainly, if very quick returns are necessary, full custom would be ruled out, and semicustom is the avenue
which we will here consider.

A survey of development times currently quoted by a range of semicustom vendors shows a considerable spread of time from receipt of
design information from the customer to delivery of prototype samples.
This is not surprising, since the following factors are involved which
influence quoted times, although the effect of some of them may not
always be clear:

1. Cell-library or masterslice design
2. The size and complexity of the required design
3. The form of the design information supplied by the customer to
   the IC manufacturer
4. The number of company or departmental interfaces in the manufacturing procedure
5. Checkback(s) to customer during the manufacturing procedure
6. The form in which the prototype samples are to be delivered
   (i.e., unscribed wafers, or bonded and packaged)

The significance of the first two factors will be readily appreciated.
All other factors being equal, a masterslice semicustom design with

commitment masks only should have a faster delivery time than a cell-library design with its full set of custom masks, and clearly, the larger the required design the longer may be its design cycle.

The remaining factors, however, may swamp the effects of items (1) and (2). For example, in item 3, if the equipment designer is capable of supplying the pattern-generation magnetic tape for the masks of his semicustom design this will be particularly advantageous; on the other hand, if only a paper design or breadboard is made available, considerable liaison and expenditure of time may be necessary. The number of different organizations involved in the manufacturing process, for example, separate mask manufacturer and IC fabrication, is also crucial, since every interface introduces potential delays and communication problems, and finally, who does what after the sample wafers come from the fabrication line is also significant.

Table 1.10 attempts to supply some ballpark figures for development times. The data of this histogram has been extracted from a wide range of commercial literature covering all forms of semicustom, both cell-library and masterslice. The lowest times may be read as small masterslice designs, with the highest times as large cell-library or other more sophisticated products.

TABLE 1.10   Histogram of Quoted Average Development Times to Prototype Samples

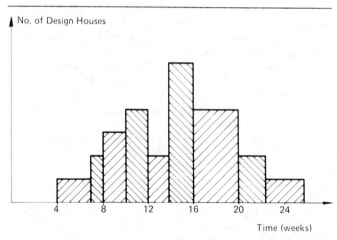

VLSI and VHSIC circuits omitted from the compilation. Full functional testing not included.
Source: Numerous commercial data brochures, 1983.

TABLE 1.11 Typical Development Schedule of
Events to Prototype Samples

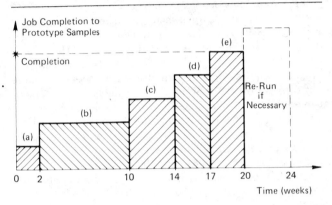

(a) Entry of design data from customer's gate or macro level partitioning. } See note in text.
(b) Mask layout design.
(c) Mask fabrication and check.
(d) Wafer fabrication.
(e) Wafer probe and package.

If we attempt to break down a quoted development time into its con-
stituents parts, then on a reasonably large circuit with a quoted 20-
week delivery we may find that the various phases require time intervals such as those illustrated in Table 1.11. The first 10 weeks shown
in this table for the IC manufacturer to absorb and use the incoming
information may be reduced by a factor of 3 or more if the customer is
capable of supplying a pattern-generation magnetic tape for his design
instead of a partitioned circuit schematic.

The question which may be asked is whether these times represent
a realistic state-of-the-art situation. Two opposing factors should be
considered: first, there is increasing experience in semicustom fabrication, and, hence, if anything these times should reduce; however,
most quoted times are for ideal situations with no unforeseen delays in
communication and across interface boundaries. A wise original equipment manufacturer would do well to at least double any times quoted to
allow for these and other contingencies; indeed six months to prototype
samples would still seem to be a ballpark figure for many semicustom
designs, with nine months to approval for equipment-quantity production.

Other constraints, in addition to time, may be present in the decision mechanism to use or not use semicustom microelectronics. *Second-sourcing* may be one, that is, can the original equipment manufacturer rely upon one source of supply for his semicustom product, or should he be in a position to send his mask set(s) to more than one fabrication line once his design is proven? If he does try this, will the products from a second fabrication line give identical performance even though the process is reputedly identical? These and other factors are part of the decisions which may constrain an original equipment manufacturer from adopting semicustom for a particular application.

Turning to the limitations of semicustom, these are generally technological ones, which are in a process of continuing evolution and improvement. Such factors as maximum size of semicustom circuit available, maximum number of input/output pins available per package, maximum operating speed available, and maximum voltage or power ratings available clearly impose a limit to the use of semicustom should the equipment manufacturer's requirements be particularly severe or unusual. However in spite of these factors, the technical advances and the applicability of semicustom techniques to the great majority of original equipment manufacturers is clear. Let us, therefore, finally consider certain other commercial factors in order to complete this general introductory survey.

1.7 COMMERCIAL ECONOMICS, COMMERCIAL SECURITY

As with the technological limitations, the economics of semicustom is an evolving situation. Here we will discuss financial factors in the framework of typical current costs, with the appreciation that the absolute values shown on any graphs may change. The general principles, however, will remain unchanged.

The breakeven points between various methods of equipment design and manufacture generally show that semicustom should be considered when the production requirements are in the thousands. Figure 1.6 gives typical target economics. Very small quantity requirements clearly are best met by off-the-shelf standard IC products; full custom is only justified for very large quantity requirements. However, it is seen that semicustom can viably span a considerable range of quantity requirements, and it is this feature which emphasizes the increasing worldwide importance of the semicustom approach. The information contained in Figure 1.6 may appear in alternative guises in literature relating to a particular form of semicustom realization, but the general picture will broadly fall into the breakeven points given herewith.

An analysis of currently quoted design house costs to prototype samples is given in Table 1.12. These costs are averaged from minimum

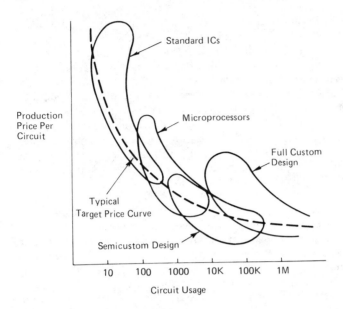

FIGURE 1.6 The economics of custom and semicustom LSI design.
(Data courtesy of Plessey Company, Caswell, England. (Note: the
price-per-circuit axis is not quantified, since prices are generally
still falling.)

and maximum development costs quoted by various IC manufacturers
for their particular range of semicustom products. As with the previous histograms of development time, small masterslice gate arrays
tend to lie to the left, with large cell-library products to the right;
also, costs are lower if the original equipment manufacturer can provide
the pattern-generation magnetic tape rather than requiring the design
house to work from a circuit schematic or breadboard.

These figures give a broad picture of quoted costs covering a range
of circuit sizes, technologies, and semicustom techniques, and cannot
be taken as other than general indicators when a particular requirement is under consideration. We can, however, attempt to break down
the overall costs of Table 1.12, to give the general trend of costs with
specific paramters.

Figure 1.7 indicates how mask costs vary with chip size, these figures being based upon an eight-mask set. In general, mask costs are
falling due to increased competition and higher throughputs, particularly the breakpoint and gradient of cost with increasing chip sizes caused
by more expensive camera requirements. Note that chip size is quoted
in mil-side-average, see introductory *List of Symbols and Definitions*;
also, the term "blowback" refers to large color reproductions on mylar
sheet of each mask, usually ×100, which may be superimposed for visual

TABLE 1.12 Histogram of Currently Quoted Average Development Costs to Prototype Samples

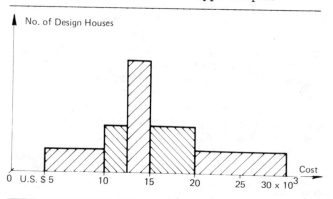

VLSI and VHSIC circuits omitted from this compilation.
Addition (repeat) design cycle $1-3K for gate array,
$2-10K for cell library.
Full functional testing not included.

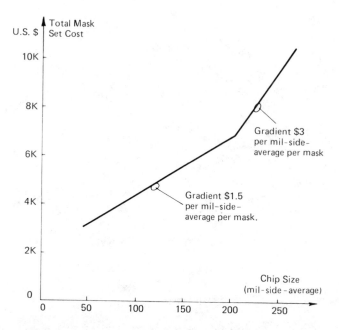

FIGURE 1.7 Mask costs for an 8-mask CMOS process, including color blowbacks and two sets of working plates.

checking purposes, and, finally, total costs are approximately proportional to the number of masks required by the particular fabrication or
custom-dedication process.

Bonding and packaging costs vary with (1) the number of pins per
package, (2) the type of package, and (3) to a lesser extent the number of circuits to be packaged. Figure 1.8 indicates typical current
figures, but again prices may be generally falling. Note that fully
automated production-line bonding and packaging as in volume-production off-the-shelf ICs is unlikely to be used for prototype and small-
quantity batches.

A final summary of current typical prototype costs for increasing
size CMOS masterslice semicustom circuits is given in Figure 1.9. The
equation for the cost graph is

Cost = $ (4.0 + 0.037n)K

where n is the number of 2-input NAND gate equivalents per chip.

Costs after acceptance of prototype samples depend upon: (1) production run quantities; (2) the technology of the semicustom circuit;
(3) other contractual arrangements between customer and IC manufacturer.

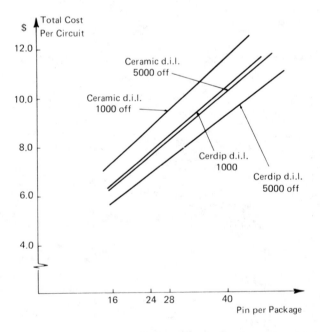

FIGURE 1.8 Typical bonding and packaging costs per circuit.

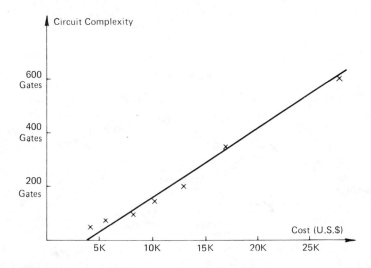

FIGURE 1.9 Development costs to 50 functional prototype CMOS circuits. (From commercial data brochure, Master Logic, Inc., Sunnyvale,
CA., 1981.)

The latter clauses may often disguise the true costs of either the
prototype samples or the production run quantities. For example,
prototype samples may be supplied below cost should there be a signed
agreement for final production orders, or alternatively, the first may
be loaded at the expense of follow-up production quantities if the latter
are indeterminate. However, following from our previous data, Figure
1.10a gives typical 1000-off costs for CMOS circuits, including bonding
and packaging, but excluding any functional testing. Figure 1.10b
indicates how different silicon technologies may affect the processed
wafer costs, but for small quantities the remaining manufacturing costs
of handling, bonding, packaging, and so on may swamp the difference
between dissimilar technologies.

Lastly, it must be appreciated that the semicustom fabrication costs
are only one part of the total which the original equipment manufacturer
(OEM) has to consider in relation to his final commercial product. What
we have not so far considered are

1. The basic system design costs and overheads of the OEM
2. The remaining hardware costs of the complete equipment
3. The complete equipment assembly and testing costs
4. OEM documentation costs, including end user documentation

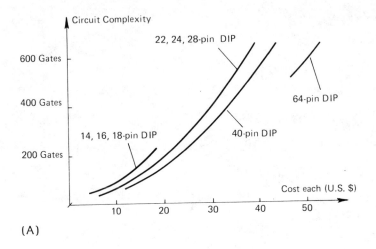

(A)

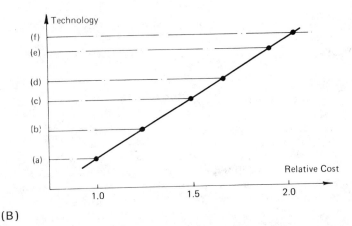

(B)

FIGURE 1.10 Semicustom costs after prototype approval. (A) CMOS production run base-line costs per package, plastic DIL packaging, 1000 off. (From Master Logic, Inc., Sunnyvale, CA.) (B) Relative manufacturing costs for fully processed wafers: (a) CMOS metal-gate; (b) nMOS silicon-gate; (c) CMOS silicon-gate; (d) ECL; (e) I^2L; (f) Low-speed TTL. (From Strategic, Inc., San Jose, CA.)

When all of these necessary costs are included, then the actual IC costs, whether off-the-shelf or semicustom, may shrink to minor financial significance.

It is this final financial picture which we cannot give herewith, since it is entirely dependent upon the particular equipment complexity, market prospects, and other unique factors. What can be done, however, is to give an illustration of the figures which an original equipment manufacturer may compile when considering costs of his new equipment. The figures following are entirely fictitious, and serve merely to illustrate the financial equations involved.

Let us consider a small piece of equipment, which may be designed and manufactured with

1. Off-the-shelf SSI/MSI/LSI packages
2. Full custom LSI, = minimum silicon area and maximum performance
3. Cell-library semicustom
4. Masterslice semicustom

We may now list first the nonrecurring (design) costs, and second the per unit production costs. These may, for example, be as follows.

Nonrecurring OEM Costs

Item	1	2	3	4
OEM circuit design + PCB and remaining hardware design	$20,000	$10,000	$10,000	$10,000
Custom or semicustom design costs to the OEM	–	150,000	20,000	15,000
OEM and end-user documentation	3,000	2,000	2,000	2,000
Total nonrecurring design costs (TDC)	$23,000	$162,000	$32,000	$27,000

Per Unit OEM Production Costs

Item	1	2	3	4
ICs plus their PCB assembly	$200	$40	$50	$60
Other hardware + assembly costs	500	500	500	500

Per Unit OEM Production Costs (Continued)

Item	1	2	3	4
Unit test	100	50	50	50
Total OEM cost per unit (PUC)	$800	$590	$600	$610

A plot of this information versus the number of units manufactured is given in Figure 1.11. This illustrates the typical cross-over of one manufacturing technique with another as the number of units manufactured varies. Whether the above costs, and hence, the actual cross-over points shown in Figure 1.11 represent any realistic piece of equipment is not important, but whatever the real-life figures may be trends as shown herewith will be observed.

The results in Figure 1.11 may of course be replotted as per unit costs versus number of units manufactured, which is obtained by dividing the y-axis values by corresponding x-axis values for each design strategy. The cross-over production point between any two strategies is given by

$$TDC(x) + N.PUC(x) = TDC(y) + N.PUC(y)$$

whence

$$N = \frac{TDC(x) - TDC(y)}{PUC(y) - PUC(x)}$$

where $TDC(x)$ and $TDC(y)$ are the total nonrecurring design costs by methods (x) and (y), respectively, $PUC(x)$ and $PUC(y)$ are the manufacturing per unit costs by methods (x) and (y), respectively, and N is the number of units manufactured.

Finally, let us briefly look at the question of commercial security of the final end product which the original equipment manufacturer sells on the open market.

An immediate advantage to the equipment manufacturer in adopting custom or semicustom technology is that it becomes harder for a competitor to copy the equipment design, since the competitor does not have access to the custom or semicustom chips. This must be viewed in contrast with, say, microprocessor-based equipment designs, where private copying of the dedication software residing in ROM is readily possible.

It is technically possible to determine the design of a given integrated circuit by careful chemical etching and microscopic examination [28, 58]. A full custom chip would pose the greatest difficulty,

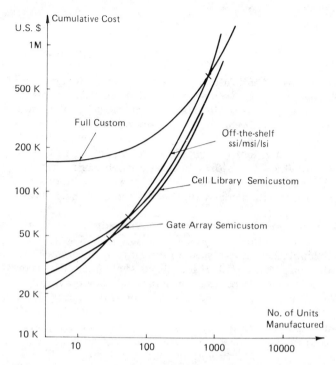

FIGURE 1.11   Total cumulative design and manufacturing costs for a
fictitious piece of equipment, with differing design strategies.

particularly if the precise fabrication technology was unknown, whereas
a masterslice semicustom chip would be considerably easier, since only
the metalization commitment need be determined. The latter statement
assumes that the uncommitted masterslice chip can be identified, and
hence only the interconnection dedication need be traced to reveal the
precise system details.

However, it is reasonable to assume that analysis and copying of a
custom or semicustom chip is not a ready or an economic proposition
for the competitor of any original equipment manufacturer to undertake—
the economics may be different if pirating of a standard off-the-shelf
volume-market product is involved; fortunately, in this text we are not
concerned with such standard-product copying. It is, therefore, safe
to say that the original equipment manufacturer enjoys considerable
commercial security if his equipment incorporates his own custom or
semicustom packages.

1.8 SUMMARY

In this chapter we have attempted to introduce the basic features, both technical and economic, of custom and semicustom large-scale-integrated techniques. We have seen that full handcrafted custom design, such as is employed in the design of volume-production microelectronic circuits, is in general so expensive in design manpower that it is not an economic proposition for an original equipment manufacturer to consider. Instead, various semicustom techniques have emerged which minimize the amount of individual design effort required to produce a dedicated circuit, and which therefore, become a much more attractive proposition for many applications. These semicustom techniques, together with programmable LSI devices will be the topic of the following chapters.

In the financial considerations which we have already covered, the conclusion which may be drawn is, that on purely financial grounds, the adoption of semicustom is unlikely to appreciably reduce the final selling cost of equipment to the end user. Some small financial saving is possible, even for very small production runs, but large overall savings are unlikely. What then is the true impact and importance of semicustom integrated circuit design? It lies outside simple financial sums, and instead is much more in the area of good engineering, design innovation, improved equipment performance, and commercial advantages. Let us conclude by listing these possible factors:

1. Enhanced and repeatable circuit performance
2. Ability to allow the system design engineer to innovate once again at gate or macro level
3. Reduced total equipment assembly size and power dissipation, with consequent possible reduced cooling requirements
4. Reduced production line assembly and test
5. Reduced inventory and documentation
6. Increased system reliability due to fewer parts
7. Increased commercial security due to difficulty of copying

Additional detailed commercial reports and other publications may also be cited [48, 59-74].

REFERENCES

1. Agajanian, A. H. *Semiconductor Devices: A Bibliography of Fabrication Technology*. Plenum Press, New York, 1976.
2. Agajanian, A. H. *MOSFET Technologies*. Plenum Press, New York, 1980.

3. AMI Staff. *MOS Integrated Circuits*. Van Nostrand, New York, 1972.
4. Barbe, D. F. (Ed.), *Very Large Scale Integration: Fundamentals and Applications*. Springer, New York, 1980.
5. Cobbold, R. S. C. *Theory and Application of Field Effect Transistors*. Wiley-Interscience, New York, 1970.
6. Fairchild, Inc. *CMOS Data Book*. 1977.
7. Fletcher, W. I. *An Engineering Approach to Digital Design*. Prentice Hall, Englewood Cliffs, NJ, 1980.
8. Glazer, A. B., and Subak-Sharpe, G. E. *Integrated Circuit Engineering*. Addison Wesley, Reading, MA, 1977.
9. Grove, A. S. *Physics and Technology of Semiconductor Devices*. John Wiley, New York, 1967.
10. Gosling, W., Townsend, W. G., and Watson, J. *Field Effect Transistors*. Wiley-Interscience, New York, 1971.
11. Hamilton, D. J., and Howard, W. G. *Basic Integrated Circuit Engineering*. McGraw-Hill, New York, 1975.
12. Hibberd, R. G. *Integrated Circuits*. McGraw-Hill, New York, 1969.
13. Mavor, J. *MOST Integrated Circuit Engineering*. Peter Peregrinus, London, 1973.
14. Mavor, J., Jack, M. A., and Denyer, P. B. *Introduction to MOS LSI Design*. Addison-Wesley, London, 1983.
15. Meyer, C. S., Lynn, D. K., and Hamilton, D. J. (Eds). *Analysis and Design of Integrated Circuits*. McGraw-Hill, New York, 1968.
16. Richman, P. *MOS Field-Effect Transistors and Integrated Circuits*. John Wiley, New York, 1973.
17. Schilling, D. L., and Belove, C. *Electronic Circuits, Discrete and Integrated*. McGraw-Hill, New York, 1979.
18. Smith, R. J. *Circuits Devices and Systems*. John Wiley, New York, 1976.
19. Smith, R. J. *Electronics: Circuits and Devices*. John Wiley, New York, 1980.
20. Streetman, B. E. *Solid State Electronic Devices*. Prentice Hall, Englewood Cliffs, NJ, 1980.
21. Taub, H., and Schilling, D. L. *Digital Integrated Electronics*. McGraw-Hill, New York, 1977.
22. Warner, R. M., and Fordemwalt, J. N. (Eds.). *Design Principles and Fabrication of Integrated Circuits*. McGraw-Hill, New York, 1965.
23. Slater, S. P., and Cox, A. M. A bipolar gate array family with a wide application/performance coverage. *Proc. WESCON Conf.*, paper 13/3, 1979.
24. Ramsey, F. R. Automation of design for uncommitted logic array. *Proc. IEEE, 17th Design Automation Conf.*, 100-107, 1980.

25. Rappaport, A. (Ed.). Semi-custom IC directory. *EDN*, 28(4): 78-218, 1983.
26. Hart, K., and Slob, A. Integrated injection logic: a new approach to LSI. *IEEE J. Solid-state Circuits*, 343-351, April 1979.
27. Berger, H. H., and Wildmann, S. K. Merged-transistor logic (MTL): a low-cost bipolar logic concept. *IEEE J. Solid-State circuits*, 340-346, October 1972.
28. Muroga, S. *VLSI System Design*. Wiley-Interscience, New York, 1982.
29. Smith, J. E. (Ed.). *Integrated Injection Logic*. John Wiley, New York, 1980.
30. Prioste, J. *MECL 10000 Macrocell Array Design Manual*. Motorola, Inc., 1979.
31. Dworsky, L. N. *Modern Transmission Line Theory and Applications*. John Wiley, New York, 1979.
32. Mead, C., and Conway, L. *Introduction to VLSI Systems*. Addison-Wesley, Reading, MA, 1980.
33. Hnatek, E. R. *A Users Handbook of Semi-conductor Memories*. John Wiley, New York, 1977.
34. Twaddell, W. Uncommitted IC logic. *EDN*, 25(7):89-98, 1980.
35. Howes, M. J., and Morgan, D. V. (Eds.). *Large-Scale Integration; Devices, Circuits and Systems*, John Wiley, New York, 1981.
36. Larson, T. L., and Downey, C. Field programmable logic devices. *Electron Engineering*, 52:37-54, January 1980.
37. Noach, K. A. H. New developments in integrated fuse logic. Mullard Technical Publication M82-0032, 1982, and *Electronic Components and Applications*, 4:111-124, February 1982.
38. Calvan, N., and Durham, S. J. Field programmable arrays. *Electronics*, 109-114, July 5, 1979, and 132-139, July 19, 1979.
39. Data I/O Corporation. *How to Survive in the Programming Jungle*. Data I/O Corp. Publication, March 1979.
40. Birkner, J. Reduce random-logic complexity. *Electronic Design*, August 16, 1978.
41. Texas Instruments. MOS programmable logic arrays. Texas Instruments Application Bulletin CA.148, 1978.
42. Wood, R. A. A high-density programmable logic array chip. *Trans. IEEE* C.28:602-608, 1979.
43. Wood, R. A., Hsieh, Y. N., Price, C. A., and Wang, P. P. An electrically alterable PLA for fast turn-around VLSI development hardware. *Trans IEEE*, J. Solid-State Circuits, 570-577, 1981.
44. Cline, R. C. A single-chip sequential logic element, *Proc. Int. Solid-State Circuits Conf.*, 204-205, 1978.
45. Rehman, M. A. Custom-integrated circuits. *Electronic Engineering*, 52:55-68, April 1980.

46. Loesch, W. Custom IC's from standard cells: a design approach. *Computer Design*, 227-232, May 1982.
47. Farina, D. E., Duffy, J. R., and Kellgren, T. L. Cell library system accommodates any degree of design experience. *Electronics*, 126-129, November 30, 1981.
48. Market Study Report. *LSI Gate Arrays*. I.C. Cost Consultants Inc., November 1979.
49. Hartmann, R. F. Design and market potential for gate arrays. *Lambda*, 1(3):55-59, 1980.
50. Roffeslen, L. Gate arrays, a user's prospectus. *Lambda*, 2(1): 32-36, 1981.
51. Walker, R. Logic arrays, technology and design. *Proc. IEEE MIDCON*, 1981. (Reprinted as Reprint No 122, LSI Logic Corp.)
52. Huffmann, G. D. Gate array logic. *EDN*, 26(19):86-96, 1981.
53. Lipp, R. Understanding gate arrays ensures wise chip selection. *EDN*, 26(19):99-104, 1981.
54. Posa, J. G. Gate arrays, a special report. *Electronics*, 145-158, September 25, 1980.
55. Kroeger, J. H., and Tozun, O. N. CAD pits semicustom chips against slices. *Electronics*, 119-123, July 3, 1980.
56. Hartmann, R. F., and Walker, R. LSI gate arrays outpace standard logic. *Electronic Design*, 107-112, March 15, 1981, and 13, 16, August 6, 1981.
57. Peters, T., and Gold, M. Logic arrays supplant discrete packages. *Electronic Design*, 63-69, December 24, 1981.
58. Business Week. How "silicon spies" get away with copying. 180-187, April 21, 1980.
59. Special issue. *VLSI Design: Problems and Tools*. *Proc. IEEE*, 70:3-190, January 1983.
60. *Dataquest Gate Array Study*. Dataquest Inc, 1981.
61. Morrow, J., and Victor, J. Looking ahead: trends and forecasts. *EDN*, 28(10):304, 1983.
62. Various papers. *Proc. International Confs. on Semi-Custom I.C's*, London, 1981, 1982, 1983.
63. Various papers. *Microelectronics Journal*, published bimonthly, 1971 onwards.
64. Various papers. *Proc. IEEE Annual Custom Integrated Circuits Confs.*, ("CICC"), 1980, 1981, 1982, 1983.
65. Various papers. *Journal of Semi-custom I.C's*, published quarterly, 1(1): September, 1983.
66. Report. *Gate Arrays Implementing LSI Technology*. Source III, distributed by Electronic Trend Publications, CA, 1982.
67. Report. *Electrically Erasible Non-Volatile Semiconductor Memories*. Electronik Centralen, distributed by IPI, Copenhagen, 1982.
68. Report. *The Impact of Custom Circuit Alternatives on Future Production Costs*. Strategic, Inc., distributed by IPI, Copenhagen, 1982.

69. Report. *The Impact of Gate Array Technology on IC Compon-ents*. Strategic, Inc., distributed by IPI, Copenhagen, 1983.
70. Report. *Standard Cell Custom IC Report*. HTE Management Resources, distributed by IPI, Copenhagen, 1982.
71. Report. *Computer Aided Engineering Workstations for the IC Industry*, Strategic, Inc., distributed by IPI, Copenhagen, 1983.
72. Krejeik, M., and Mash, S. (Eds.). *Semicustom IC Yearbook*. Benn Publications, U.K., 1983.
73. Various papers. *European Semiconductor Design and Production*. published quarterly, 1979 onwards.
74. Supplement. *Semicustom Circuits Special Report*. *Electronic Times*, November 18, 1983.

2
Basic Semiconductor Technologies:
A Review

2.1 INTRODUCTION

It is arguably not necessary for a digital equipment designer who may be utilizing semicustom microelectronic circuits in his products to be familiar with semiconductor physics and device technology. One may make this claim when off-the-shelf microelectronic packages are used, and indeed many excellent texts are available which contain a minimum of device physics [1-6]. Nevertheless, the equipment designer usually has some knowledge of the inner workings of the devices he is employing. The hallmark of a good engineer is an inquiring mind, and therefore some overlap of his knowledge into the device area is invariably present.

Hence in the custom and semicustom field, the original equipment designer (customer) will have or will acquire a general appreciation of semiconductor technology and fabrication details, although obviously not in the depth which is the province of the manufacturer (vendor) of integrated circuits (ICs). We include this chapter on basic semiconductor technologies so that this text may be more widely useful. What we will not attempt to do, however, is to cover any intimate details of crystal growth, fabrication line machinery, mask making, the effect of varying parameters on yield, and so on; these indeed may be considered areas of expertise beyond the scope of this present text.

We will first consider the main semiconductor technologies used in semicustom devices, and finally attempt to compare and contrast their relative advantages and performance capabilities. Greater depth of information may be found in the references cited during this general discourse.

The principal division in present-day semiconductor physics and technology is between bipolar and unipolar devices. In the former, both types of charge carrier, that is, both negatively charged electrons and positively charged holes, take part in the semiconductor device action; in the latter only one type of carrier, either negative electrons or positive holes, constitutes the major conduction stream of the device.* Within each division we have several major semiconductor technology families, as indicated in Tables 2.1 and 2.2.

It should be noted that in the bipolar families of Table 2.1, emitter-coupled logic (ECL) is also a form of current-mode logic (CML), since nonsaturating transistor currents are steered from one path to another during the circuit action (see Sec. 2.2.7). However, we have here retained the terminology current-mode logic for circuit configurations in which preset currents are turned on and off in association with other nonsaturating switching transistors.

It may also be noted that in unipolar families the original terminology metal-oxide semiconductor (MOS) was specific to earlier devices which employed a metal gate with silicon dioxide insulation; the subsequent buried polysilicone gate devices have however still retained the designation MOS, even though this is strictly inaccurate. Further, one may remark that the terminology "transistor," which originally was an abbreviation taken from "transfer-resistor" and which correctly represented the two-port circuit performance of the common-base junction transistor, is inappropriate for unipolar MOS devices; however we appear to be stuck with these everyday designations which are a feature of expediency rather than technological accuracy.

In the following sections, we will review the general features and characteristics of the families shown in Tables 2.1 and 2.2 in sufficient detail for most original-equipment-manufacture (OEM) design engineers involved in semicustom digital design activities. We will not become involved in such depth as may be necessary for custom design at the transistor level, particularly when analog or hybrid requirements are involved, or if device geometries are being shrunk to micron dimensions; the latter problems are still strongly the concern of the professional IC designer, until such time as device performances are fully

---

*Recall that conduction by means of positively charged holes in a given direction is actually the movement of a drift of negatively charged electrons in the opposite direction. However, because the mechanism of charge movement of the so-called holes is dissimilar to that of simple electron conduction, it is usual to regard them as a movement of positive charge, with specific mobility and other characteristics.

TABLE 2.1 The Major Semiconductor Technology Families[a]

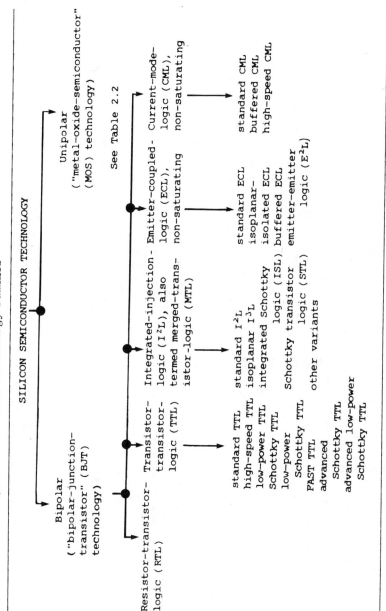

TABLE 2.2 The Major MOS Semiconductor Families[a]

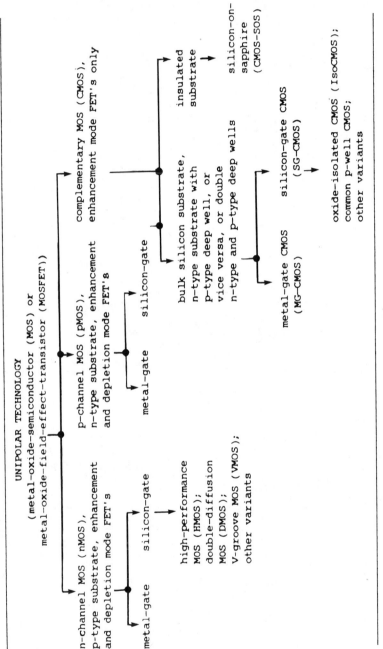

UNIPOLAR TECHNOLOGY
(metal-oxide-semiconductor (MOS) or
metal-oxide-field-effect-transistor (MOSFET))

n-channel MOS (nMOS),
p-type substrate, enhancement
and depletion mode FET's

metal-gate   silicon-gate

high-performance
MOS (HMOS);
double-diffusion
MOS (DMOS);
V-groove MOS (VMOS);
other variants

p-channel MOS (pMOS),
n-type substrate, enhancement
and depletion mode FET's

metal-gate   silicon-gate

complementary MOS (CMOS),
enhancement mode FET's only

bulk silicon substrate,
n-type substrate with
p-type deep well, or
vice versa, or double
n-type and p-type deep wells

metal-gate CMOS   silicon-gate CMOS
(MG-CMOS)   (SG-CMOS)

oxide-isolated CMOS (IsoCMOS);
common p-well CMOS;
other variants

insulated
substrate

silicon-on-
sapphire
(CMOS-SOS)

[a]Silicon-gate technology generally provides a higher performance capability than metal-gate and
therefore is becoming predominant.

quantified and broadcast. Finally, we will omit the resistor-transistor
logic (RTL) class of Table 2.1, since this is no longer a technology in
widespread use. Certainly as far as the semicustom area is concerned,
the alternative bipolar and MOS technologies are currently more generally and increasingly representative of the semicustom arena.

Further general details of all these microelectronic families may be
found in many publications [7-17].

2.2 BIPOLAR TTL, I²L, ECL, AND OTHERS

Common to all bipolar families is the bipolar junction transistor, operating either (1) in a saturating switching mode, or (2) in faster circuit configurations in some nonsaturating mode, usually with reduced
voltage swings in order to minimize the effects of circuit capacitance
and device storage.

While both polarities of bipolar junction transistors, *pnp* and *npn*,
are possible and indeed available, in the following discussions we will
confine ourselves to *npn* devices, being the usual polarity for all
present-day custom and semicustom activities, since this generally
provides a higher device performance (gain, switching speed, etc.)
than *pnp* counterparts. The two device polarities are of course complementary to each other; *p*-type semiconductor material in one is *n*-type
in the other and vice versa, while all polarities of voltage and current
are reversed one to the other. It is perhaps unfortunate that the basic
terminal action of transistors is best introduced around the supply polarities of the *pnp* transistor, using conventional (positive) directions
of current flow, but we will not avail ourselves here with this facility.

In addition, we will confine all our following considerations to
silicon-based devices, although the future will certainly see the expansion of microelectronics based upon gallium-arsenide and other alternative semiconductor compounds. We will see that, almost invariably,
bipolar fabrication is based upon a *n*-type epitaxial layer grown upon
a *p*-type substrate, the only exception of note being the collector-
diffusion-isolation (CDI) fabrication technique, which will be discussed
later.

2.2.1 Basic *npn* Transistor Device Action

The basic *npn* transistor device action in TTL, integrated-injection logic
(I²L), and all other bipolar transistor families is as follows; the different bipolar families involve detailed differences in device construction
or circuit use, which we will subsequently consider.

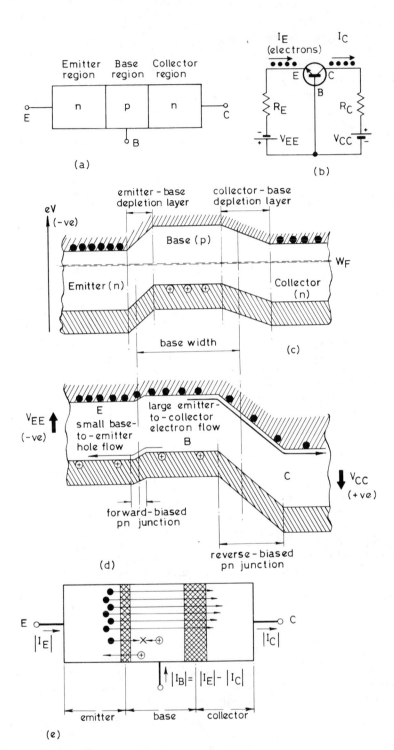

Emitter
region
Base
region
Collector
region

n   p   n

E   C

B

(a)

I_E (electrons) I_C

E   C

B

R_E R_C

V_{EE} V_{CC}

(b)

emitter-base
depletion layer

collector-base
depletion layer

eV
(-ve)

Base (p)

W_F

Emitter (n)

Collector (n)

base width

(c)

V_{EE}
(-ve)

E   small base-
to-emitter
hole flow

large emitter-
to-collector
electron flow

B

C V_{CC}
(+ve)

forward-biased
pn junction

reverse-biased
pn junction

(d)

E   C

$|I_E|$ $|I_C|$

X—⊕
⊕

$|I_B| = |I_E| - |I_C|$

emitter   base   collector

(e)

48

Consider the simplified *npn* transistor structure shown in Figure 2.1a. The three adjoining regions within the regular crystalline silicon structure are:

1. The emitter region, which is doped to produce a *n*-type semi-conductor, that is, an excess number of electrons are present compared with the number within a perfect quadravalent crystal lattice structure (see Appendix A)
2. The base region, of narrow effective width, and doped so as to produce *p*-type semiconductor properties
3. The collector region, which like the emitter is doped to exhibit *n*-type semiconductor properties

It is possible to make the overall structure symmetrical, whereby the emitter and collector regions are identical and interchangeable, but for optimum performance the emitter region is normally more highly doped than the collector (see Sec. 2.2.2). At first sight the *npn* transistor may appear to be two back-to-back *pn* junction diodes with a common *p* region. Indeed, if the base region was very wide, then the device would generally behave as two separate diodes, but when the base has a very narrow effective width, which may be as small as 0.5 μm, then the following familiar transistor action results.

Consider the supply polarities shown in Figure 2.1b. The V_{EE} supply is a forward-bias polarity across the emitter-base *pn* junction, (see Appendix A, Fig. A.7), and therefore will result in a current flow across this junction. The V_{CC} supply, however, is such as to reverse-bias the collector-base *pn* junction, and hence by itself would not cause any appreciable collector current flow.

Perfect transistor action requires that all the electron carriers in-jected by V_{EE} from the emitter region into the base region flow across the base and appear in the output circuit, being "collected" by the applied collector supply voltage V_{CC}, that is ideally

$$I_C = I_E, \text{ with } I_B = 0$$

FIGURE 2.1   The *npn* bipolar junction transitor (BJT): (a) the sche-matic of the *npn* structure; note the base region is made very narrow; (b) symbol, showing the polarity of external dc supplies for normal circuit action, and the direction of electron carrier flow in emitter and collector; (c) the electron energy band model in thermal equilibrium, with no external applied voltages; (d) the electron energy band model in the active mode, V_{EE} and V_{CC} present; (e) a final simplified picture, showing some electron-hole recombination in the base region, but ig-noring further effects.

This requires two principal features to be present in the base region, namely: (1) that the effective base region shall be narrow, as already mentioned, and (2) that the concentration of acceptor impurities in the base p region shall be lower than the concentration of donor impurities in the emitter region, so that there are relatively few hole carriers available compared with the electron carrier population. Under these circumstances, V_{EE} will cause a large concentration of electron carriers to be injected into the base region; a small number of hole carriers will also flow across into the emitter region, and some injected electrons will combine with holes in the base region, this hole carrier action constituting the base current I_B of the transistor.

However, the majority of electrons injected into the base will drift through this narrow base region until they come under the influence of the base-collector voltage gradient, where they will immediately be swept up and collected by the collector potential. The donor doping of the collector region is somewhat of a compromise, requiring a low donor impurity level in order to provide a wide depletion region and a narrow effective base width, while on the other hand requiring a high donor impurity level in order to match the emitter-to-base potential gradient, and make the overall emitter-to-collector potential small.

The electron energy band model for this action is shown in Figure 2.1c and d. The final static (or dc) transistor circuit action is such that:

$$I_C = h_{FB} I_E \qquad\qquad (2.1)$$

where h_{FB} is the large-signal (or dc) current gain of the transistor in common-base configuration

$$I_B = (1 - h_{FB}) I_E \qquad\qquad (2.2)$$

V_{BE} = typically 0.6 V for silicon

V_{CE} = typically 0.2 V for silicon

and where h_{FB} may be as high as 0.995 or more under exceptional circumstances.* These current equations assume that the injected emitter current dominates all other carrier effects.

This simplified explanation and the transistor configuration shown in Figure 2.1 is the common-base circuit action. Of more practical

---

*h_{FB} is a parameter which is current-, voltage-, and temperature-dependent as well as device-dependent, and may have a range of values approaching but never exceeding 1.0.

significance in virtually all semicustom applications is the transistor reconnected in common-emitter mode, as shown in Figure 2.2a. The base may now be considered as the input terminal of the transistor, the collector remaining the output terminal as before.

It should be appreciated that the action within the transistor remains unchanged for a given set of applied voltages, irrespective of which transistor terminals are considered to be the circuit "input" and "output." Hence, Eqs. (2.1) and (2.2) still apply. However, from Eq. (2.2) we have

$$I_E = \frac{I_B}{1 - h_{FB}}$$

whence

$$I_C = h_{FB} \left(\frac{I_B}{1 - h_{FB}} \right)$$

$$= h_{FE} I_B \qquad\qquad (2.3)$$

where

$$h_{FE} = \frac{h_{FB}}{1 - h_{FB}} = \text{the large-signal (or dc) current gain of the transistor in common-emitter configuration.}$$

Eq. (2.3) gives the output current I_C in terms of what is now considered to be the input current I_B, and since $h_{FB} \to 1.0$ it is clear that $h_{FE} \gg 1.0$, giving the familiar large value of current gain of the common-emitter configuration.

It is entirely possible to discuss the fundamental bipolar transistor action directly in the common-emitter configuration, without basing it upon the common-base configuration. Because some readers may prefer this approach, we briefly outline it below, so that we may have a broader view of the transistor action.

Consider the components of current present in the base circuit under active conditions. We may consider three components, namely:

1. $I_{B(1)}$, which corresponds to the number of hole carriers which flow out of the p-type base to the n-type emitter region across the forward-biased base-emitter junction (see Appendix A, Fig. A.7e)

2. $I_{B(2)}$, which corresponds to the number of hole carriers which have to be supplied by V_{BB} to replace those which have recombined in the base region with electrons flowing in across the forward-biased emitter-base junction

3. $I_{B(3)}$, which is the result of thermally generated electron-hole pairs in the collector region, which if generated within one diffusion length of the collector-base depletion layer will result in the holes being swept across into the base by the reverse-bias collector-base voltage

This last component of current is the temperature-dependent collector-base leakage current, which we have not previously considered. Indeed we need not consider it further, except to note that in this common-emitter configuration its presence tends to make the base region acquire a slightly more positive potential than would otherwise be the case, resulting in a higher electron emission from the emitter into the base region, and finally a higher total collector current.*

However because the emitter region is a *heavily doped n* region and the base region is a *very lightly doped p* region, the number of available electron carriers in the emitter region far exceeds the number of available hole carriers in the base region. Thus the forward-biasing of the base-emitter junction which allows current $I_{B(1)}$ to take place will simultaneously allow a vastly greater number of electrons to enter the base region from the emitter. Some will combine with holes, to form the base current component $I_{B(2)}$ above, but the vast majority will remain available to diffuse across the base and be collected by the base-collector voltage. Hence, we establish, as before, that $|I_C|$ is considerably greater than $|I_B|$, with $|I_C| \simeq |I_E|$. These considerations are illustrated in Figure 2.2, which should be read in conjunction with the previous Figure.

The principal overall terminal characteristics of the transistor are as illustrated in Figure 2.3. Exact values on the axis of each graph will vary between transistors, particularly the I_C/I_B and h_{FE} parameters. It may be noted that:

1. The I_B/V_{BE} characteristic is a normal *pn*-junction diode-type characteristic, with V_{BE} never exceeding, say, 0.6 V for silicon.
2. The $I_B = 0$ intercept on the I_C/I_B characteristic is the collector-to-emitter leakage current I_{CEO}, which itself is approximately given by $I_{CEO} = h_{FE} I_{CBO}$.
3. The h_{FE}/I_C characteristic shows a reduction in h_{FE} value at very low values of I_C due to a higher percentage recombination of the emitter electrons in the base region, and also at high current levels due to the swamping of the base region with emitter electrons and the resulting higher percentage which then spill into the base circuit.

---

*This is the feature that can cause thermal runaway and excessive collector currents if temperatures are very high and circuit stability is inadequate.

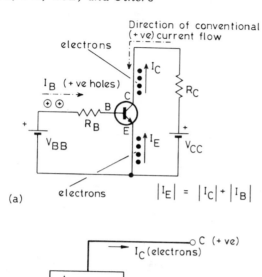

(a)

$$|I_E| = |I_C| + |I_B|$$

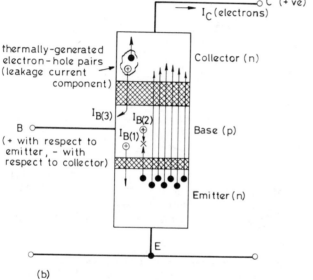

(b)

FIGURE 2.2   The *npn* bipolar junction transistor in common-emitter configuration: (a) polarity of external dc supplies for normal circuit action, and direction of carrier flows; (b) the three components of base current (see text), with the major current conduction being electrons from the *n*-type emitter to collector.

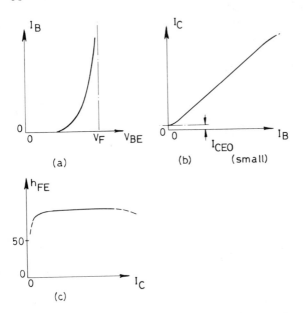

FIGURE 2.3 The principal terminal characteristics of the *npn* bipolar
junction transistor: (a) base-emitter characteristics; (b) forward current transfer characteristics, V_{CE} constant; (c) forward large-signal
gain characteristics, V_{CE} constant.

This is the outline action of all bipolar transistors as used in semicustom and other applications. The above description is specific to
the transistor in its active and nonsaturated region of use; to conclude
let us define the three possible regions which may be encountered in
practical circuits:

1. The cut-off region of application, in which all transistor currents are minimal; this normally implies that V_{BE} as well as V_{CB}
   is reversed-biased
2. The active region, in which V_{BE} is forward-biased, but V_{CB}
   remains reverse-biased as shown in Figure 2.1d
3. The saturated region, in which V_{BE} remains forward-biased,
   but the collector voltage has fallen due to external resistance
   in the collector circuit such that V_{CB} is no longer reverse-
   biased; no increase in collector current can now occur, and both
   pn junctions are forward-biased

Bipolar transistors used in a simple switching mode therefore are in
either state 1 or state 3, as indicated in Figure 2.4. For further

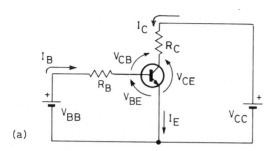

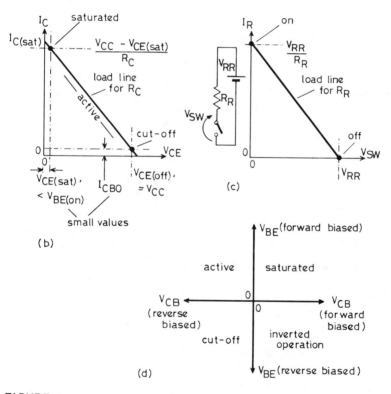

FIGURE 2.4 The three main regions of bipolar transistor action:
(a) the common-emitter circuit with collector load R_C, and direction of
conventional (positive) current flows; (b) I_C/V_{CE} loadline, with cut-
off at I_{CBO} assuming zero base-to-emitter voltage, and saturation at
$V_{CE(sat)}$; (c) in comparison, a perfect on/off switch; (d) summary,
inverted operation seldom encountered.

physical details, see Pierret [18], Neudeck [19, 20], and other texts
[15, 16, 22-24].

2.2.2 Epitaxial Transistor Fabrication

The fabrication of bipolar transistors and the detailed circuit configurations vary considerably, as will now be considered. We will, however,
only consider integrated circuit transistor fabrication using epitaxial
techniques, and not that of individual discrete transistors such as may
be used in power applications.
     Figure 2.5 shows a basic integrated circuit *npn* transistor structure.
The starting point is a high-resistivity *p*-type bulk substrate on which
the following principal fabrication steps are performed:

   1. A deep well of n^+, termed the buried layer is first diffused into
       the substrate, which will eventually provide an internal low-
       resistance path for the collector output current.
   2. A lightly doped *n*-type epitaxial layer is then grown over the
       whole surface, which is then divided into the individual isolated
       transistor areas by diffusing p^+ boundaries through the epitaxial
       layer to the *p*-type substrate.
   3. A lightly doped *p*-type area is then diffused into the *n*-type
       epitaxial area to form the base of the transistor.
   4. A highly doped n^+ area is then diffused into 3 to form the cen-
       tral emitter area, and also a similar n^+ area is simultaneously
       diffused into the previous *n*-type epitaxial region to form the
       final ohmic collector connection to the epitaxial collector region.
   5. Final contacts are made to emitter, base, and collector regions
       by first covering the surface with a thin metal deposit, with
       subsequent selective etching to leave the required inter-
       connections.

     It should be noted that the *p*-type substrate is always made the
most negative potential of the assembly. Hence, we have a *pn* junction
border around each *n*-type collector area, with the *p* region negative
with respect to the *n* region. This is a reverse-bias *pn* junction diode
situation (see Appendix A, Fig. A.7d), giving the so-called *diode isolation* or *junction isolation* between devices on the same chip. It should
also be noted that the "layers" shown in Figure 2.5 are extremely thin,
and hence the emitter → base → collector currents may be considered
to flow entirely vertically in this figure, through a very thin base
region layer.
     The principal masking, etching, and other steps required in the
above process are indicated in Figure 2.6 and listed below:

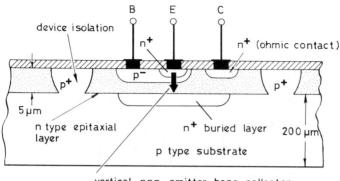

vertical npn emitter-base-collector

FIGURE 2.5   The basic structure of an epitaxial integrated-circuit *npn* transistor with reverse-biased *pn* junction device isolation. (Note: vertical proportions are exaggerated for clarity.)

1.   Silicon dioxide is grown over the whole surface area.
2.   Cut windows through step 1 surface.
3.   Diffuse deep n^+ buried layer.
4.   Remove remaining surface oxide.
5.   Grow n-type silicon over whole surface by epitaxial growth, to subsequently become collector regions.
6.   Grow silicon dioxide layer on top of step 5 surface.
7.   Cut windows for junction-diode device isolation.
8.   Diffuse p^+ region through to substrate to form the isolation moats in the epitaxial layer.
9.   Regrow silicon dioxide surface layer.
10.   Cut windows for base implant.
11.   Diffuse weak p region into the epitaxial layer to form base region.
12.   Repeat step 9.
13.   Cut windows for emitter and collector connection implants.
14.   Diffuse shallow n^+ regions to form emitter region and ohmic collector connection.
15.   Repeat step 9.
16.   Cut windows for contacts.
17.   Cover whole surface with metal.
18.   Final metalization pattern mask, followed by etching away of all unwanted metal.
19.   Final surface protection and other connection processes.

Resistors are made by the same procedures up to step 12 to form *p* type areas in the epitaxial layer, low-resistance ohmic connections

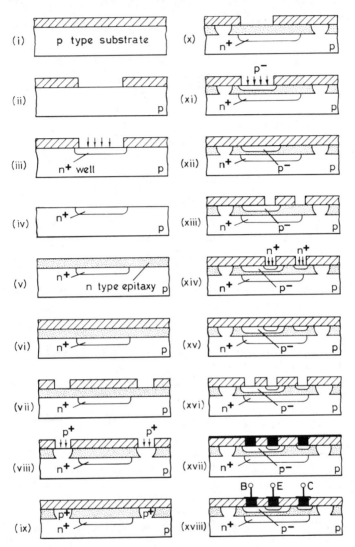

FIGURE 2.6   The individual steps in the fabrication of Figure 2.5.
See text for details of each step.

then being made to these separate appropriately dimensioned p-type
areas to complete these ohmic resistors. In total, as many as 13 or
more masks per complete mask set and more than 20 process stages
may be involved in bipolar fabrication. Further details may be found
in more specialized references [13, 14, 25-27].

2.2.3 Schottky TTL

Referring back to Table 2.1, the principal families of bipolar transistor circuits are listed as RTL, TTL, I^2L, ECL, and CML. Resistor-
transistor logic (RTL), will not be considered in this text, since as
previously mentioned it does not constitute a family generally employed
in the custom or semicustom arena. The others, however, are well
represented.

The standard TTL logic circuit configuration as used in standard
off-the-shelf small- and medium-scale integration (SSI and MSI) is illustrated in Figure 2.7a. For on-chip use in dedicated circumstances the
output may be considerably simplified from the general-purpose push-
pull output configuration shown here. The main disadvantages of this
configuration stem from the saturation mode of the switching transistors,
which slows the gate propagation times, together with the power per
gate taken from the supply rail. The latter places a severe limit on
the number of gates that can be provided per chip without running into
chip dissipation problems.

To reduce these adverse effects, enhanced low-power and Schottky
variants have been developed. These are particularly relevant for the
custom and semicustom area. The low-power variants are in general
higher resistance versions of standard TTL configurations; the Schottky
variants, however, incorporate Schottky diodes in order to eliminate
saturation and improve gate propagation times.

The Schottky diode is a rectifying junction which forms between
aluminum and n-type silicon.* The characteristics of such a diode are
that it conducts at about half the value of the p silicon/n silicon junction diode, and has negligible storage time; on the other hand it has
a poor reverse breakdown voltage which precludes its use as a rectifying device in its own right. However, if a Schottky diode is formed
across the collector-to-base of a transistor, it will conduct before the
collector-to-base pn junction can become forward-biased, and hence
it will provide a path to prevent collector saturation. Figure 2.7b illustrates this concept. It may be appreciated that when the collector
potential drops below that of the base as the former approaches

---

*With heavily doped n^+ silicon, however, an ohmic contact is made
with aluminum, as in fabrication stage 14 previously detailed.

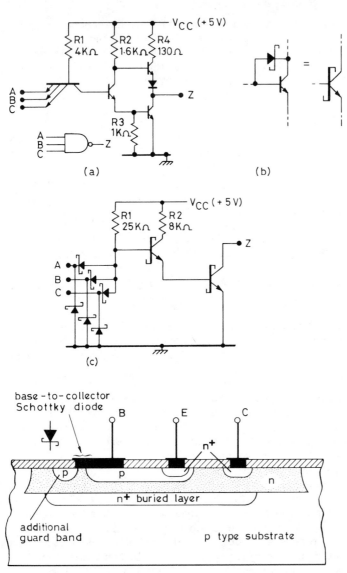

FIGURE 2.7 Bipolar TTL circuits: (a) the standard discrete TTL
NAND logic gate configuration; (b) the Schottky diode antisaturation
addition, and combined diode/transistor symbol; (c) typical low-power
Schottky TTL circuit, open-circuit collector output; (d) the formation
of a Schottky diode between base and collector.

saturation, then the Schottky diode will begin to conduct, which we may then regard as either allowing excess positive base input current to be diverted from the transistor base to the collector, or providing a path for additional collector electrons to flow back to the base supply circuit. Either consideration precludes the collector potential from dropping further to its otherwise $V_{CE(sat)}$ value.

Figure 2.7c illustrates a low-power Schottky TTL NAND gate, as may be used on-chip. Figure 2.7d indicates how the base connection is allowed to overlap the epitaxial n-type collector region, thus forming the Schottky clamping diode from the base terminal to the n-type collector region.

Further TTL variants specific to particular IC manufacturers may be found, and doubtless additional variants will continue to appear. However, turning to the alternative main bipolar families (see Table 2.1), let us now consider the integrated-injection logic (I^2L) family, also sometimes termed merged-transistor logic (MTL). We will use the former designation herewith.

2.2.4 Integrated-Injection Logic (I^2L)

Integrated-injection logic (I^2L or IIL), is a particular family of bipolar logic configurations whose fabrication steps are similar to those which we have already described. The circuit action, however, is completely dissimilar.

The basic I^2L gate is shown in Figure 2.8a. It consists of a *pnp* transistor Tl, termed the "injector," which is used both as an injector of base current into the *npn* transistor T2 of the same gate, and as a current-source load for a transistor collector output of a preceding gate. If this input base current is feeding into T2, it allows each collector to act as a sink (low resistance to 0V). However, if this current source is deflected away by the presence of a sink connection at the gate input, then the T2 output collectors can no longer act as sinks, and effectively become open-circuit paths. Note that the collector outputs can never act as current sources to supply further circuits connected to them. Thus the basic I^2L gate is a single-input, multiple-output configuration, unlike more familiar logic gates which have multiple inputs (gate fan-in) and a single output.

The assembly of I^2L gates is generally as shown in Figures 2.8b and c. Note that we have revised the position of the injector transistor Tl in these figures to agree more closely with the actual fabrication (see Figure 2.9), but the principle has not been altered. In Figure 2.8b, each I^2L gate is acting as an invertor (NOT) gate, while in Figure 2.8c the action is:

Inputs		Outputs		
A	B	$\overline{A}$	$\overline{B}$	Z
Low	Low	Not sunk	Not sunk	Not sunk (high)
Low	High	Not sunk	Sunk (low)	Sunk (low)
High	Low	Sunk (low)	Not sunk	Sunk (low)
High	High	Sunk (low)	Sunk (low)	Sunk (low)

which if "not sunk" (high) is regarded as logic 1, gives us the 1-level output function $Z = \overline{A} + \overline{B}$.

Several features of this circuit action which are in contrast to TTL circuit action may be noted:

1. The logic voltage swing at gate inputs and outputs is only the difference between $V_{CE(sat)}$ when a collector output is acting as a sink, and $V_{BE(on)}$ when the injector current I_B is not being sunk.

2. Constant current I_B switches between the base input of the multiple-collector *npn* transistor and the output collector of a preceding gate, which makes the gate power consumption approximately constant.

3. The collector currents are merely the injector currents I_B, there being no collector resistors R_C to V_{CC} as in the TTL case, and hence total power consumption per gate is extremely small.

Against these advantages must be weighed the interfacing problem between the very low logic-level swings per gate and the outside world, the latter normally requiring 0–5 V logic levels. However, it is perfectly straightforward to fabricate both I^{2}L and TTL circuit configurations on the same chip, and therefore interfacing is not a fundamentally difficult procedure.

The I^{2}L fabrication remains an epitaxial fabrication, involving the *n*-type epitaxial layer on *p*-type substrate as developed in Figures 2.6 items i–v inclusive. However, the *pnp* injector transistor is fabricated as a lateral transistor in contrast to the vertical *npn* transistor, and also one injector emitter region with a common resistor to +V_{CC} can serve many *npn* output transistors. This is illustrated in Figure 2.9, which may be read in concert with the previous TTL fabrication diagrams. Notice that the multiple collectors are now within the *p*-type base region and the emitter is the epitaxial layer, which is opposite to that of the normal TTL action of Figure 2.5.

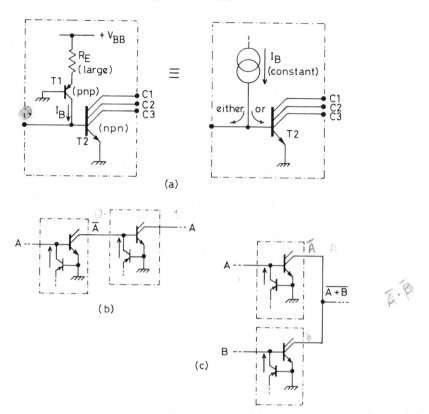

FIGURE 2.8   I^{2}L logic gate configurations: (a) basic gate structure;
(b) cascade of logic inversions; (c) wired collector outputs, function
Z = $\overline{A + B}$.

I^{2}L offers high packing density with considerably reduced power
consumption in comparison with standard TTL. Its speed performance,
however, is generally not as good as Schottky TTL, but may be improved by further refinements and alternative detailed structures.
   One improvement is to replace the deep p-type diffusion moats surrounding each transistor island (see Figure 2.6 item viii and 2.9) by
a moat of silicon dioxide, which has the result of diminishing the depletion layer capacitance which is otherwise present.* This oxide

---

*The depletion layer at any reverse-biased pn junction acts as a
capacitance in the circuit; hence the pn-diode isolation border acts as
a capacitance connected to the circuit.

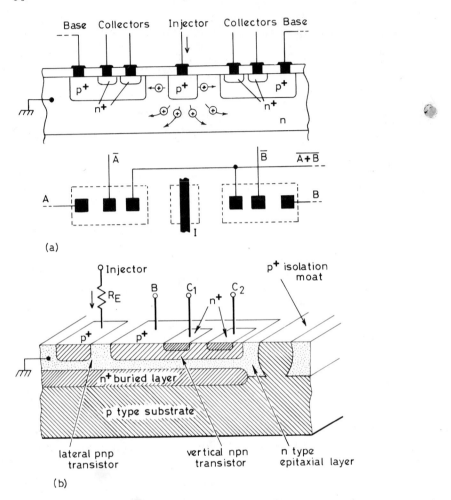

FIGURE 2.9 The physical fabrication of I^2L gates: (a) schematic of Fig. 2.8c, exaggerated scale; (b) epitaxial fabrication. (Note: isoplanar fabrication similar except silicon dioxide device isolation instead of p^+.

isolation technique is usually referred to as "isoplanar," and therefore the terminology isoplanar I^2L, or I^3L, may be encountered. Additional advantages are that the contact diffusions may now extend out to the isolation border (or conversely the borders may be brought in), which reduces device size and eases mask alignment tolerances.

 I^3L may show a three- or fourfold improvement in speed compared with normal I^2L, but requires additional mask and fabrication steps.

Further bipolar developments which provide enhanced circuit performance may be found. We will now briefly consider these. In Section 2.6 we will make a final comparison of all the various bipolar and MOS families which we will have considered.

2.2.5 Integrated-Schottky Logic (ISL)

Integrated-Schottky logic is a development of the basic I^2L technology, which takes advantage of the best features of both I^2L and Shottky TTL to provide very fast and compact circuit configurations. Like I^2L, two transistors are combined to form a single-input, multiple-output configuration, but unlike I^2L we now find:

1. The *npn* transistor collector is back in the *n*-type epitaxial region, with Schottky diodes to provide the necessary isolation between the several collector output points.
2. The current source to switch on the *npn* multiple-output transistor comes via the input electrode instead of from the internal merged *pnp* constant-current injector source.
3. The *pnp* transistor is now appropriately doped so as to act as a collector clamp on the *npn* transistor to prevent V_{CE} saturation.

The circuit configuration of the ISL gate is shown in Figure 2.10a, with the epitaxial structure given in Figure 2.10b. Very fast switching speeds are now possible due to the lower capacitance and storage effects associated with the Schottky-isolated outputs, but the presence of the Schottky diodes in the outputs reduces each output voltage swing by some 200 mV compared with the voltage swings of the normal I^2L process.

2.2.6 Schottky-Transistor Logic (STL)

The last of these closely related families of I^2L circuits which we will consider is the Schottky-transistor logic configuration. While this conveniently follows on from the previous ISL discussions, the final STL circuit configuration closely resembles the Schottky TTL circuits discussed earlier. However, the STL and ISL circuits do retain a close affinity to each other, since many of the stages of fabrication are identical.

Figure 2.11 shows the circuit configuration and structure of a STL logic gate. The *pnp* transistor clamp of the ISL gate has now been replaced by a Schottky diode, similar to the Schottky TTL arrangement given in Figure 2.7, while the collector output isolation remains as in the ISL gate.

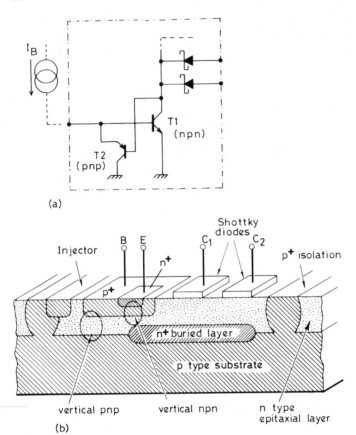

FIGURE 2.10   Details of ISL gates: (a) basic circuit configuration
(two or more Schottky-diode-isolated outputs); (b) epitaxial fabrication.

The current supply for the base of transistor Tl is normally through
a polysilicon resistor rather than the previous lateral *pnp* transistor.
The final result is somewhat more complex to fabricate than ISL gates,
typically requiring 12 mask stages compared with 10 for ISL, but the
speed performance may be twice as fast as ISL. Hence, as we will detail more fully later, in increasing speed capability we have I^2L, I^3L,
ISL, and STL, all of which are characterized by low power dissipation
per gate, and therefore are appropriate for large-scale integration
(LSI) and semicustom activities.

2.2.7 Emitter-Coupled Logic (ECL)

Turning now to an entirely new family of bipolar logic circuits, we have the very high speed emitter-coupled logic (ECL) family. This family currently provides the fastest available logic circuits, but this speed performance has the penalty of high power dissipation per gate.

Figure 2.12a shows the basic circuit configuration of an ECL gate. It consists of a two-transistor emitter-coupled circuit, with transistors operating either in the cut-off or the active region, but never in the saturated mode. With a constant base potential V_{REF} on transistor

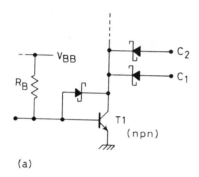

(a)

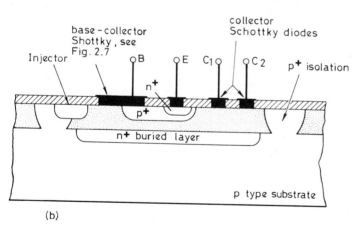

(b)

FIGURE 2.11   Details of STL gates: (a) basic circuit configuration (typically four Schottky diode outputs); (b) epitaxial fabrication.

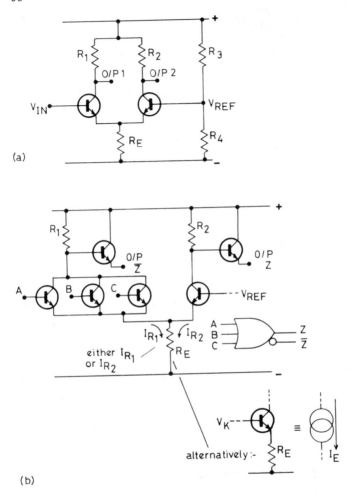

FIGURE 2.12   The ECL gate: (a) basic ECL configuration; (b) 3-input OR/NOR circuit arrangement.

T_2, transistor T_1 is off and T_2 is conducting when input V_{IN} is less than V_{REF}, switching to T_1 conducting/T_2 off when V_{IN} is greater than V_{REF}. Thus the emitter current through resistor R_E flows through one side or the other depending upon the value of V_{IN}, provided the V_{IN} voltages lie either side of V_{REF}. Since resistors R_1 and R_2 are roughly the same value (not necessarily exactly balanced), approximately constant current is taken by the whole circuit, irrespective of the logic output.

For logic purposes, two or more transistors are placed in parallel, as shown in Figure 2.12b. Emitter-follower outputs for O/P1 and O/P2 are provided in order that the gate output logic levels shall be compatible with gate input voltage requirements, and also to provide low output impedance driving capabilities. Note that these output emitter-follower transistors are always in their active region.

The values of resistors R_1 and R_2 are chosen such that the final emitter-follower output voltages lie approximately symmetrically above and below V_{REF}. From Figure 2.12b it will be apparent that output Z realizes the OR of the inputs to the left-hand half of the circuit, while output $\overline{Z}$ is the complement. The ECL gate, therefore, is an OR/NOR configuration. Further, it is directly possible to electrically tie together the emitter-follower outputs from two or more ECL gates, any one gate output being high (logic 1) causing the common connection to be high, thus giving wired-OR capability. This is otherwise known as emitter-dotting or implied-OR, and is easily implemented by metalization (dotting) between emitter output points.

Thus this family of circuits provides a powerful logic capability with the merits of (1) very high switching speeds due to the elimination of transistor saturation with its resultant storage time constraints, (2) approximately constant current taken from the supply rails, thus eliminating power rail spikes, (3) ability to operate over a wide temperature range, and (4) good drive capability to charge and discharge circuit capacitances. Against these merits must be weighed the disadvantages of (i) very high power dissipation per gate in comparison with other technologies, thus limiting the number of gates per chip, (ii) a small output voltage swing between logic 0 and logic 1, which is not compatible with the majority of final off-the-chip requirements, and (iii) usually two separate on-chip power supplies are provided, rather than the potential-divider mechanism to provide V_{REF} shown in the considerably simplified diagram of Figure 2.12a.

Early ECL fabrication methods followed the epitaxial constructions such as we have already considered. However, since the emphasis with ECL has been very strongly towards faster speeds, the isoplanar techniques for gate isolation which we have previously seen applied to I^2L circuits are usually employed. This has the further merit of increasing the possible gate packing densities. Further improvements in packing densities and isoplanar isolation may be seen in the later isoplanar II and isoplanar S developments of particular company products. Figure 2.13 illustrates these general ECL fabrication methods.

Further variants on ECL circuit configurations may be found, including:

1. Buffered ECL, which has additional circuit components to give improved tolerance to supply voltage and other variations

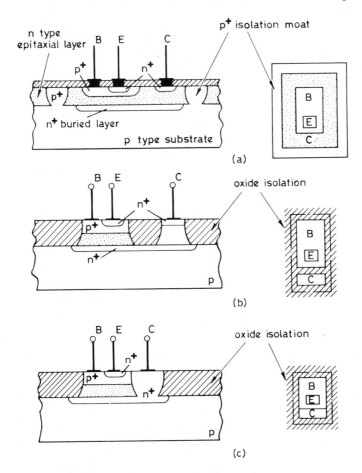

FIGURE 2.13   Emitter-coupled logic fabrication: (a) basic *pn* junction
diode isolated transistor construction, approximately 3×10^3 μm^2 per
transistor (c.f., Figure. 2.5); (b) replacement of *pn* junction isolation
by isoplanar silicon-dioxide isolation, approximately 1.5×10^3 μm^2 per
transistor; (c) further isoplanar developments, approximately 800 μm^2
per transistor. (note: resistors in the ECL circuits are epitaxial areas
of appropriate geometry and resistivity.)

2. Emitter-emitter logic ECL (E²L), which is specific for driving 50 Ω transmission line loads
3. Lower-power ECL, with or without emitter-follower outputs, running on a reduced power supply voltage and providing slower gate speeds

It seems likely, however, that the main use for ECL in the future will only be where ultimate speed of operation is essential; other bipolar or MOS technologies will cater adequately for all other areas. Certainly in the semicustom area it is not anticipated that ECL will play a significant part, except possibly for mainframe computer requirements, where speed may have to match computing specifications, and for state-of-the-art signal processing requirements.

2.2.8 Current-Mode Logic (CML)

The previous ECL circuits may themselves be termed "current-mode" circuits since their circuit action is essentially the steering of an approximately constant current from one part of the gate circuit to another. However, we customarily reserve the term for the particular family of bipolar circuits which we will now consider.

Figure 2.14a gives the basic current-mode logic (CML) configuration for a simple INVERTER gate. With V_{IN} at logic 0 (low voltage) the switching transistor T_1 is off, and the output is at logic 1 ($\simeq V_{CC}$). With V_{IN} at logic 1 (high voltage), T_1 is switched on, and the output drops to logic 0. However, the current in the "on" state is controlled by the preset current Ip in the emitter, and is limited so that T_1 does not reach saturation. Hence, T_1 is only operating in either the off or the active mode, and never in the saturated mode.

The current source in the emitter is controlled by a multiple-emitter transistor T_2 working in its inverse mode, that is, each available emitter acts as an entirely separate isolated collector, with the true collector acting as a common emitter for all the multiple "collectors." The particular base-"collector" biasing of T_2 provides a current-mirror action, that is, a precisely defined current flow in the dedicated "collector," and all other collectors mirror this particular value of current when called upon to pass current. It is incorrect to regard the emitter circuit of the switching transistor T_1 as a constant-current source, because this normally implies that a constant value of current flows under all circuit conditions; here we have a preset rather than a constant current, which either flows when T_1 conducts or does not flow when T_1 is off. Thus, unlike the ECL constant-current circuit, we are switching currents completely on and off under normal logic action, and therefore total gate dissipation is, on average, much reduced. Furthermore, the one current-source circuit can serve several logic gates by virtue of its isolated "collector" terminals.

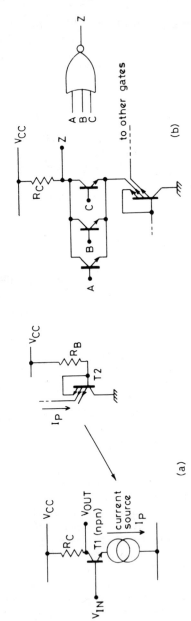

(b)

(a)

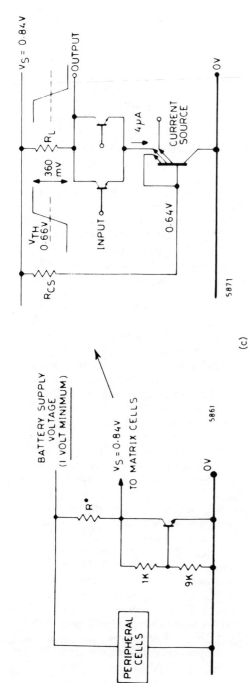

FIGURE 2.14   The CML gate: (a) basic inverter configuration;   (b) 3-input NOR gate;   (c) possible on-chip voltages for battery operation.   (Details courtesy Ferranti, Ltd., Manchester, England.)

Extension to multiple-input gates readily follows, as shown in Figure 2.14b. It will be seen that the basic gate action is a NOR configuration. Precise voltage and current levels may be varied considerably, since the basic circuit configuration has wide supply voltage tolerance. Generally by increasing V_{CC} and reducing R_C a faster switching operation is obtained at the expense of average power dissipation, but at the other extreme extremely low on-chip current and voltage levels can be utilized, for example as shown in Figure 2.14c. Hence, different families of CML gates may be found, with various tradeoffs between speed, power, and chip area. Emitter-follower output buffers may also be provided at each gate output in order to increase the gate output drive capabilities.

A requirement of the circuit configurations shown in Figure 2.14 is that the multiple-emitter transistor shall have an efficient operation in its inverse mode. This is particularly provided by collector-diffusion isolation (CDI) fabrication technology, and therefore current-mode logic is invariably associated with CDI technology. Both have been extensively pursued and marketed by Ferranti Ltd., in the United Kingdom with their associated company Interdesign Inc., as we will consider further in Chapter 4.

The CDI fabrication process is still an epitaxial process, but unlike all the processes so far considered does not involve a n-type epitaxial layer on the surface of the p-type substrate. Instead a p-type epitaxial layer is laid down, which will constitute the final transistor base regions, n^+ diffusion into this epitaxial layer finally forming the collector and isolation regions. Note that we have now covered three methods of isolation for planar epitaxial devices, namely separate reverse-biased pn-junction diode isolation (see Fig. 2.5), isoplanar oxide isolation (see Fig. 2.13), and as will be shown below combined collector-diffusion isolation.

Figure 2.15 shows the general fabrication details of the CDI processes. The first stages are similar to those detailed in Figure 2.6 items i—iv, giving a buried n^+-type layer, but the following stage involves the growth of a p-type epitaxial layer rather than the n-type layer of Figure 2.6 item v. Shallow diffusion of n^+ into the p-type eptiaxial layer then provides base and resistor areas, with deep n^+ diffusion through the epitaxial layer providing the combined collector and isolation regions. It will be seen that the collector is a heavily doped n-type region, unlike the lightly doped collectors of TTL fabrication; this gives the technology the good inverse-mode transistor operation necessary for the current sources, but at the expense of diffusion times of electrons across the base region. A particular feature of the CDI process is that thin epitaxial thicknesses (typically 2 μm) and shallow diffusion depths (typically 1 μm) are employed, giving a very narrow base width and hence a good operating speed.

Further features of the CDI process include the ability to use n^+ diffusions as a supply connection on the chip and to distribute V_{CC} to

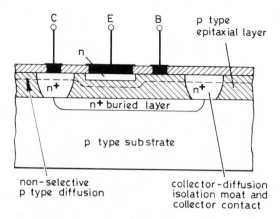

C   E   B   p type epitaxial layer

n

n+

n+ buried layer

p type substrate

non-selective
p type diffusion

collector-diffusion
isolation moat and
collector contact

FIGURE 2.15 Fabrication details of the collector-diffusion isolation (CDI) process.

the collector resistors. At the same time the p-type substrate can be used as the ground (0 V) return from all circuits.

This completes our survey of the principal variants of the bipolar families of circuits, and their main fabrication details. Further details of masking, etching, diffusion, and other processes such as ion implantation may be found in many publications, together with a greater depth of treatment of device characteristics than we have summarized herewith [7, 8, 13, 14, 16, 25-30].

2.3 MOS TECHNOLOGY

In constrast to the bipolar transistor considered in the preceding sections, whose action depended upon interrelated hole and electron movements in the three regions of the transistor, field-effect transistors (FETs) are the active device elements in nMOS technology, and also in pMOS and CMOS which we will subsequently consider.

The field-effect transistor is a device in which the control of the current between two terminals, the "source" and the "drain," is by means of an electric field applied across the conducting channel. The voltage to establish this field is applied to the third device terminal, the "gate," the gate-to-source voltage V_{GS} being the controlling influence. By varying the gate voltage, the effective cross-sectional area for conduction between source and drain or the number of available carriers may be varied, and hence the resulting current flow between these two points may be controlled. In the limit, the path for conduction may be cleared of carriers, thereby effectively open-

circuiting the source-to-drain path. The source-to-drain current involves only one type of charge carrier in this conduction path, either
electron carriers or hole carriers, unlike our previous bipolar transistors; field-effect transistors, therefore, are *unipolar*, not bipolar,
semiconductor devices. However, we will still encounter *pn* junction
structures in FET technology, which may be deliberately employed as
diode isolation boundaries between devices.

There are two principal types of field-effect transistors as indicated
in Table 2.3 and below: (1) the junction-gate FET, sometimes termed
JFET; (2) the insulated-gate FET, sometimes termed IGFET, MOSFET,
or MOST. In the latter, the input impedance of the gate terminal is
effectively infinite, and zero gate current is present under static conditions. In the former, however, a small value of gate current is present. In this text we will concentrate entirely on the insulated-gate
FET, since this is the most widely used in all digital circuit applications
and in all semicustom activities. We will use the basic abbreviation
FET where required without further qualification, rather than IGFET
or other alternatives.

2.3.1 Basic Device Characteristics

Although we will finally be largely concerned with enhancement-mode
insulated-gate FETs for reasons we will discuss later, let us begin with
an outline consideration of the depletion-mode metal-gate transistor,
which may be either a *n*- or a *p*-channel device.

Figure 2.16a shows the basic construction of a *n*-channel depletion-
mode FET. The substrate is *p*-type silicon in which are diffused n^+
source and drain areas, with a *n*-type conducting channel between
them. Above the conducting channel is the metal-gate electrode, insulated from the conducting channel by a thin silicon dioxide layer.

When no connection is applied to the gate, conduction between the
source and drain can take place through the *n*-type conducting channel,
the exact value depending upon the drain-to-source voltage and the
device geometry and doping levels. The current flow is a majority-
carrier electron flow from source to drain, assuming the drain terminal
is positive with respect to the source. The *p* substrate is made the
least-positive potential, usually the same as the source potential, and
hence the *n*-type conducting channel is at some positive voltage gradient with respect to the substrate, giving reverse-biased *pn* junction
isolation between channel and substrate. The current flow between
source and drain is therefore confined to the channel area, its value
being I_{DSS} under zero gate voltage conditions.

If we consider the effect of applying a negative potential to the gate,
making V_{GS} negative, then the gate/gate-oxide/conducting channel
will act as a capacitor, with a net negative charge on the gate and a

TABLE 2.3   The Principal Divisions of Field-Effect Transistors

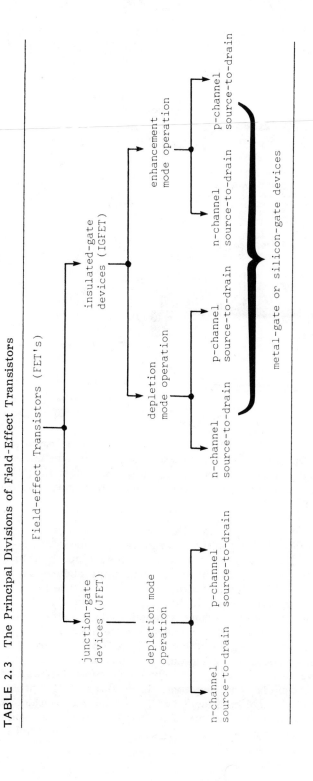

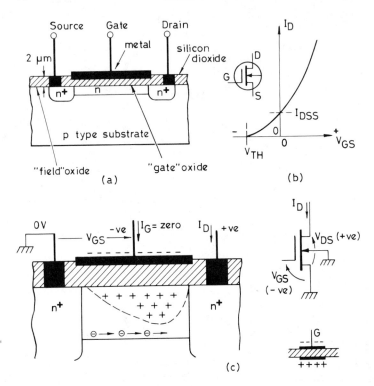

FIGURE 2.16   The n-channel metal-gate depletion-mode FET: (a) idealized nMOS structure, not to scale; (b) circuit symbol, and drain current I_D vs. gate-to-source voltage V_{GS} characteristic, V_{DS} constant; (c) enlarged vertical cross-section with gate negative with respect to source and substrate.

corresponding equal positive charge being induced on the opposite "plate" of this capacitor, namely along the conducting channel. The valency electron population in the conducting channel is, therefore, depleted. This is illustrated in Figure 2.16c. The precise physical details of the formation of the charge distribution in the conduction channel may be found in more detailed semiconductor texts [9, 15-23], but will generally result in a charge distribution as illustrated. In effect, the negative gate potential has reduced the cross-sectional area available for source-to-drain current flow, and therefore reduced the current for the same drain-to-source voltage conditions. If V_{GS} is made increasingly more negative, there comes a point at which the conducting channel is "pinched off," and no path remains for the source-to-drain electron flow. This is the negative threshold voltage V_{TH} shown in Figure 2.16b. When V_{GS} is made positive, additional negative charges are induced in the conducting path, and the resulting source-to-drain electron flow is augmented.

Thus, the gate structure is basically a capacitor structure with a very thin silicon dioxide dielectric, the control of the charges on the "plates" of this capacitor controlling the effective conduction channel for source-to-drain current. The p-channel FET of comparable structure operates in a very similar manner, except that all dopings and voltages are reversed, and the current flow is majority-carrier holes rather than electrons from source to drain.

This type of FET is termed "depletion-mode" because it is conducting when V_{GS} is zero and requires V_{GS} to be reverse-biased to deplete the conduction channel and reduce I_D to zero. Therefore, V_{GS} has to have an opposite polarity to V_{DS} in order to cut off the device, which is a serious disadvantage in many practical applications, particularly digital applications. A more satisfactory situation is present when the gate cut-off (threshold) voltage V_{TH} is the same polarity as the drain-to-source voltage supply, and $I_D = 0$ when V_{GS} is zero. To achieve this we require enhancement-mode characteristics, as will now be described.

Figure 2.17a illustrates the general structure of an enhancement-mode metal-gate FET. In comparison with the depletion-mode device of Figure 2.16a it will be observed that there is no longer a diffused n-type conduction channel joining the source and drain areas, and hence with zero applied field there will be no conduction between source and drain.

If the gate is reverse-biased (V_{GS} negative for p-type substrates), this merely increases the positive charge density in the p-type substrate below the gate. However, when the gate is forward-biased, the holes in this substrate region will be repelled, and electrons from the source and drain regions will begin to spread into the region under the attraction of the positive gate potential. Thus the positive charge on the gate electrode is mirrored by a corresponding negative charge in the substrate region. As the positive gate potential is increased, there will eventually be the situation where electrons will be present all across the surface region from source to drain, at which point the surface region is said to be "inverted" since the original hole carriers at the surface have been displaced by electrons. Hence the gate generates the n-type surface inversion channel in the p-type substrate, and electron conduction between source and drain can now take place.

The gate voltage above which conduction can just take place is the gate threshold voltage V_{TH} as shown in Figure 2.17b. Increasing V_{GS} above this critical voltage V_{TH} enhances the electron density in and depth of the surface inversion channel, thus increasing the drain current for any given drain-to-source voltage V_{DS}.

This simple explanation of the formation of the n-type surface inversion channel in the p-type substrate may be additionally substantiated by consideration of the electron energy bands within the substrate, as we will briefly indicate. Figure 2.18 shows the effect of

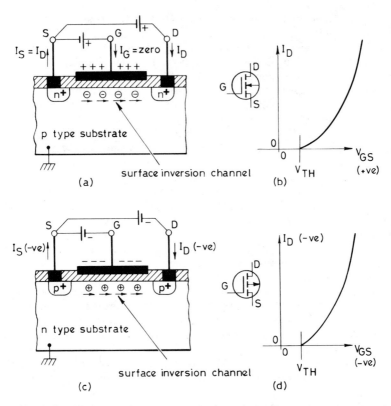

FIGURE 2.17   The metal-gate enhancement-mode FET: (a) enhancement n-channel FET; (b) circuit symbol, and I_D/V_{GS} characteristic,
V_{DS} constant; (c) enhancement p-channel FET; (d) circuit symbol,
and I_D/V_{GS} characteristics, V_{DS} constant.

the electron energies in the substrate with varying grid-to-source
voltage V_{GS}.* The p-type material has an available electron energy
level close to the valency band, into which electrons may jump, leaving
behind holes in the valency band for conduction purposes (see Appendix
A). The Fermi level E_F, that is the electron energy value at which
there is a probability of 0.5 (50%) of finding an electron, moves from

---

*We should more correctly use the gate-to-substrate voltage in this
development, but we are here continuing to assume that both source
and substrate are at 0 V, as previously. Additionally, there may
be voltage gradients in the substrate near the top surface, which
we will also ignore in this general consideration.

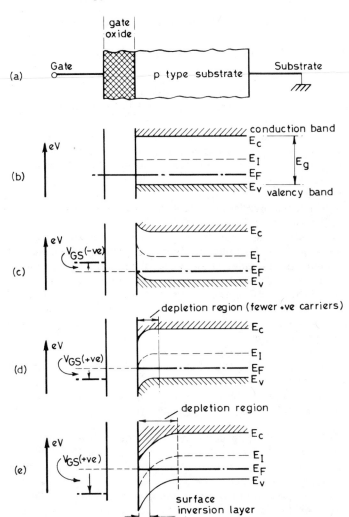

FIGURE 2.18 The energy band concept of an ideal MOS capacitor,
with the substrate at 0 V: (a) the capacitor presence in an insulated-
gate FET; (b) the situation with zero gate field, flat-band conditions;
(c) V_{GS} negative, E_v approaches or overlaps E_F, accumulation conditions; (d) V_{GS} positive, E_I approaches E_F, depletion conditions but
not yet inversion; (e) V_{GS} more positive, E_I overlaps E_F, inversion
conditions.

the middle of the forbidden energy gap E_I where it would lie for intrinsic (undoped) semiconductor material, to some value nearer E_V (see also Appendix A, Fig. A.6d). Consider the developments indicated in Figure 2.18, which are as follows.

1. Condition b: zero applied field from the gate electrode, normal p-type valency band, conduction band and Fermi level throughout the substrate.
2. Condition c: gate made negative with respect to the substrate, energy bands bend near the surface as shown, except for the Fermi energy level which is a constant reference for the whole of the bulk substrate. The gap between E_V and E_F is reduced, allowing more electrons to escape from the valency band and more holes to become available. This effect is identical to more heavily doping the p-region.
3. Condition d: gate made positive with respect to the substrate, energy bands now bend as shown, increasing the gap between E_V and E_F near the surface. Fewer electrons can now escape near the surface, and less holes become available. This is identical to more lightly doping the p region. When the energy bands bend such that E_I (the middle of the energy gap) just meets E_F at the surface, then the surface layer has just become intrinsic, and is no longer p-type.
4. Condition e: the increasing gate voltage causes the surface layer to behave identically to n-type material, since E_F is now nearer to the conduction band E_c than the valency band E_V, that is the intrinsic level midgap voltage E_I is now below E_F, which is the characteristic of n-type material.

Further details and the controlling equations for this MOS capacitor action may be found [7, 9, 15, 21]. Note that with increasing positive V_{GS}, the depletion layer thickness increases, with the n-type inversion region being eventually established.

Reverting back to Figure 2.17, the p-channel enhancement-mode FET will be seen to be identical in construction to the n-channel device which we have just considered, except that all dopings and voltage polarities are reversed. As shown in Figure 2.17c, a p-type surface inversion channel is induced in the n-type substrate when $|V_{GS}|$ is greater than the threshold voltage value, giving similar enhancement I_D/V_{GS} characteristics as in the n-channel case.

All field-effect transistors conform to this basic mode of operation, although as we will soon see there are many practical methods of fabrication. However, there are essentially only four categories of FET, whose explicit circuit symbols are given in Figure 2.19. Note that with the depletion-mode variants, where there is an implanted conduction channel between source and drain, then the circuit symbol contains a

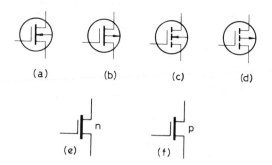

FIGURE 2.19 Summary of the insulated-gate FET symbols: (a) depletion-mode *n* channel; (b) depletion-mode *p* channel; (c) enhancement-mode *n* channel; (d) enhancement-mode *p* channel; (e) and (f) simplified symbols for *n*- and *p*-channel devices, depletion or enhancement, metal-gate or silicon-gate.

solid bar between the source and drain electrodes; in the enhancement cases where no implanted channel exists, the symbol between source and drain is broken. However, for many circuit applications it is adequate to use the simplified symbols shown in Figure 2.19e and f.

The most common equations associated with field-effect transistors are:

(i) $C_{ox} = \dfrac{\xi_0 \xi_r A}{x}$

$(2.4a)$

where

C_{ox} = the capacitance of the physical gate substrate capacitor (F)

ξ_0 = absolute permittivity (Fm^{-1})

ξ_r = relative permittivity of the gate oxide

A = gate area (m^2)

x = gate oxide thickness (m)

This may be given in per unit area, namely

$C_O = \dfrac{\xi_0 \xi_r}{x}$ Fm^2

$(2.4b)$

(ii) $\quad I_D = \frac{W}{L} \mu_N C_O \left\{ (V_{GS} - V_{TH})V_{DS} - \frac{V_{DS}^2}{2} \right\}$ $\qquad$ (2.5)

where

$\quad I_D \qquad\qquad\qquad$ = drain current (A)

$\quad W \qquad\qquad\qquad$ = width of gate (m)

$\quad L \qquad\qquad\qquad$ = length of gate (m)

$\quad \mu_N \qquad\qquad\qquad$ = drift mobility of channel electrons ($m^2 V^{-1} s^{-1}$)

$\quad C_O \qquad\qquad\qquad$ = gate capacitance per unit area (Fm^{-2})

$\quad V_{GS}, V_{TH}, V_{DS}$ = gate-to-source, threshold, and drain-to-source voltages, as previously (V)

The above equation is for n-channel devices; for p-channel devices, μ_N is replaced by μ_p, the drift mobility of channel holes. Further, $\mu_N C_O$ (or $\mu_p C_O$) may be referred to as the process gain factor K', (AV^{-2}), and in turn W/L K' may be considered as a single parameter β, termed the transistor gain factor. Hence

$$ I_D = \beta \left\{ (V_{GS} - V_{TH})V_{DS} - \frac{V_{DS}^2}{2} \right\} \qquad (2.6) $$

It will be appreciated that because the mobility μ_N for electrons is higher than that for holes, β for n-channel devices will be higher than that for p-channel devices.

(iii) $\quad I_{D(sat)} = \beta \left\{ \frac{(V_{GS} - V_{TH})^2}{2} \right\}$ $\qquad$ (2.7)

where $I_{D(sat)}$ is the saturated drain current (A) for increasing V_{DS}, given by the maximum value of Eq. (2.6).

(iv) $\quad V_{DS(sat)} = V_{GS} - V_{TH}$ $\qquad\qquad\qquad\qquad$ (2.8)

where $V_{DS(sat)}$ is the value of V_{DS} (V) for given values of V_{GS} and V_{TH} where I_D just reaches $I_{D(sat)}$, obtainable from Eq. (2.6).

(v) $\quad g_m = \beta V_{DS}$ $\qquad\qquad\qquad\qquad\qquad\qquad\qquad$ (2.9) where g_m is the mutual conductance (transconductance) of the transistor below saturation, (AV^{-1}), obtained by differentiation of Eq. (2.6).

(vi) $g_{m(sat)} = \beta V_{DS(sat)}$ (2.10)

where $g_{m(sat)}$ is mutual conductance at or above saturation.

(vii) $R_{ON} = \left\{ g_{m(sat)} \right\}^{-1}$ (2.11)

$= \left\{ \beta(V_{GS} - V_{TH}) \right\}^{-1}$

where R_{ON} is the resistance of the "on" source-to-drain conduction channel (Ω), when $V_{DS} \leqslant (V_{GS} - V_{TH})$ (i.e. when the device is hard on).

(viii) $V_{TH} = V_{FB} + 2\psi_B + \gamma(2\psi_B)^{1/2}$ (2.12)

where V_{TH} is threshold voltage, as before (V), V_{FB} is transistor flat-band voltage (V), which is a voltage dependent upon the characteristics illustrated in Figure 2.18,

$$\psi_B = \frac{kT}{q} \ell n \left(\frac{N_A}{n_I} \right)$$

that is, a parameter dependent upon the accepter level concentration N_A in the substrate and the free electron concentration n_I of intrinsic silicon (assuming n-channel device), and

$$\gamma = \frac{(2\xi_0 \xi_r q N_A)^{1/2}}{C_O}$$

that is, a further parameter dependent upon the acceptor level concentration together with the gate capacitance parameters.

The value of the threshold voltage may be roughly taken as

$V_{TH} \simeq 0.2 V_{DD}$ for enhancement devices $\Big\}$

(2.13)

$V_{TH} \simeq -0.8 V_{DD}$ for depletion devices $\Big\}$

where V_{DD} is the dc drain supply voltage as previously.

All of the above equations should be taken as approximate, since further device parameters and ambient conditions are involved. For additional information, see more detailed references [7, 9, 16, 21].

2.3.2 Basic MOS Circuits

We have summarized the MOS device equations in greater detail than we considered for bipolar transistors because MOS technology is becoming increasingly dominant for custom and semicustom applications, except where specific bipolar performance is required. However OEM designers will not generally be required to become involved in MOS physics, and therefore even the above summary may be for background information only.

Knowledge of basic MOS circuit configurations, however, will be required. Again we will concentrate here mainly upon switching circuits rather than analog, since digital applications currently dominate the custom and semicustom market, with analog currently remaining a more specialized design and market area. However, see Section 2.3.3, p. 99.

The fundamental MOS switching circuits are shown in Figures 2.20a and b. Assuming we are employing enhancement-mode devices, then with V_{GS} at 0 V, the device is nonconducting, and the output voltage lies on the x axis of the loadline plot, at $|V_{DS}| = |V_{DD}|$. With V_{GS} forward-biased greater than its threshold value, drain current flows; with $|V_{GS}|$ approaching $|V_{DD}|$ the device will be in saturation, exhibiting an "on" drain-to-source resistance R_{ON} whose value depends upon the device construction.

The output voltage V_{DS} is now controlled by the potential divider R_D and R_{ON}, being given by

$$I_D = \frac{V_{DD}}{R_D + R_{ON}}$$

(2.14)

$$V_{DS(on)} = \frac{V_{DD} R_{ON}}{R_D + R_{ON}}$$

Clearly, if $V_{DS(on)}$ is to be low, ideally 0 V for a switching circuit, then $R_{ON} \ll R_D$. R_{ON}, however, may be of the order of $K\Omega$ for small geometry devices, and hence R_D is required to be large, possible tens of $K\Omega$.

Large-value resistors are very area-consuming in microelectronic form, and it is preferable to adopt an alternative strategy in order to achieve high ohmic resistance. Since a FET behaves as a resistance between source and drain, it follows that a FET, appropriately

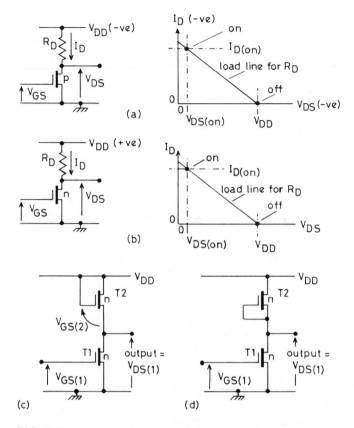

FIGURE 2.20 Basic FET switching configuration: (a) simple p-channel inverter circuit; (b) simple n-channel inverter circuit; (c) simple n-channel circuit with enhancement-mode n-channel load; (d) as (c) but with depletion-mode n-channel load.

dimensioned and biased, can be used in place of a normal integrated circuit resistor. Hence, the "loadmost" concept, as shown in Figure 2.20c.

The dimensions of the loadmost T_2 will be dissimilar to that of the switching transistor T_1, so that its "on" resistance is much higher than the resistance R_{ON} of T_1. However, if the loadmost is itself an enhancement transistor, then when the output voltage $V_{DS(1)}$ is low (ideally 0 V, = logic 0), the gate-to-source voltage $V_{GS(2)}$ of T_2 is high and the loadmost is behaving as a predetermined load resistance, but when the output voltage rises from 0 V towards V_{DD} (= logic 1), then V_{GS} of T_2 will decrease. It will now be appreciated that it is impossible for the output $V_{DS(1)}$ to reach V_{DD}, as the loadmost transistor ceases to conduct when the circuit output is $|V_{TH}|$ below $|V_{DD}|$; hence, as the output voltage rises from logic 0, the enhancement loadmost will gradually cease to conduct, becoming an increasingly large resistor as $V_{GS(2)}$ falls towards its threshold value V_{TH}.

This is an intolerable circuit action for most purposes. The alternative is to use a depletion transistor as the loadmost device T_2, with a fixed gate-to-source voltage $V_{GS(2)}$ of 0 V. This is illustrated in Figure 2.20d. T_2 now has fixed biasing conditions, which allow it to behave as an approximately constant fixed resistor R_D of appropriately large value, and allowing the output voltage to approach V_{DD}, controlled by Eq. (2.14).

The equivalents of the n-channel circuits shown in Figure 2.20c and d are possible in p-channel form. However, due to the higher mobility of the electrons in n-channel devices compared with p-channel, the former are preferable for most practical applications, and represent the widest current practice.*

Figure 2.21 illustrates circuit configurations for n-MOS NAND and NOR logic gates. In all cases the depletion-mode transistor behaves as an appropriate value resistance. If it is desired that the logic 0 output voltage of all circuits shall all have roughly the same value, then it is necessary for the silicon area geometries of the various devices to be scaled in an appropriate manner.

Taking the 3-input NAND gate of Figure 2.21a, in order for the logic 0 output voltage level of this circuit to be similar to that of, say, the INVERTER gate of Figure 2.20d, it is necessary for the "on" series resistance of the three switching transistors in the former to be similar to the "on" resistance of the single switching transistor in the latter, assuming the loadmost transistors retain the same geometry. Within the same fabrication details, the "on" resistance is controlled by the width W of the conduction channel and its length L, that is the aspect ratio W/L, being a higher resistance for increasing L, and a decreasing resistance for increasing W. Thus, for the NAND gate of Figure 2.21a we may modify the aspect ratios of the three transistors in cascade such that they behave in total the same as the single-switching transistor of Figure 2.20d.

The aspect ratio W/L is thus a measure of the conductance of the device, or, conversely, L/W represents its resistance. By adjusting the aspect ratio of the geometries of the switching transistors and the loadmost transistor, we may arrange the resistance ratios of load and

---

*Historically it initially proved easier to make stable p-channel devices than n channel. Hence, early MOS logic circuits tended to be the inferior-performance p-type, until such time as increasing fabrication expertise allowed reliable n-channel technology to be perfected [7-9].

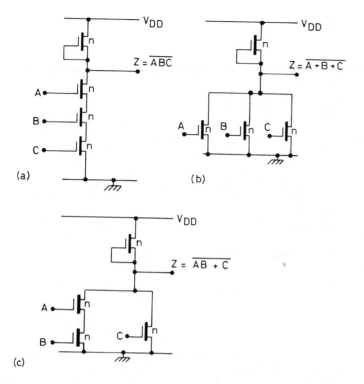

FIGURE 2.21 Simple nMOS logic circuits ("ratioed" circuits):
(a) 3-input NAND gate; (b) 3-input NOR gate; (c) composite gate.

switching transistors in any configuration to be whatever ratios are
desired for optimum switching. Indeed, the aspect ratio W/L is the
major and unique parameter in MOS circuit design, being quite unlike
any other parameter which is present in bipolar technology. Its presence may have already been noted in the several equations detailed in
the preceding section, and in total it allows the detailed on-silicon
designer to adjust the current levels and also the speed of the circuits
in a way which is not available in any other technology.

For the simple INVERTER gate given in Figure 2.20d, the aspect
ratio of the single switching transistor may be, say, six times the
aspect ratio of the loadmost transistor, with possible values such as:
switching transistor T_1: W = 24 µm, L = 4 µm; loadmost transistor T_2:
W = 8 µm, L = 8 µm. This will ensure that the logic 0 output voltage is
lower than V_{TH} of the enhancement transistors, enabling the output to
drive subsequent gates with an appropriate noise margin. It will not,
however, be as low as the logic 0 output level of the CMOS circuits we

will shortly consider. For the more complex logic circuits, appropriately different aspect ratios may be chosen.

The speed of operation of MOS gates depends very largely upon the
ability of the gate output to charge and discharge the circuit capacitance which is present on the output; we may always consider the gate
inputs as infinite dc resistance, and it is the output charging and discharging time constants which dominate. However, from our previous
considerations, it will be appreciated that a gate output impedance in
the logic 0 state is always less than the output impedance in the logic 1
state; the former is a low sink resistance to 0 V, but the latter is the
higher source resistance to V_{DD} via the loadmost resistance. Therefore, it will be the t_{OFF} performance, that is the time for the output
to recover towards V_{DD} (logic 1), rather than t_{ON} which will dictate
the maximum operating speeds.

Figure 2.22a illustrates the circuit conditions in a simple INVERTER
circuit when switching between the two logic levels. In the case of
t_{OFF} it will be the channel resistance of the loadmost transistor T_2
which controls the charging time-constant, T_1 being nonconducting;
in the case of t_{ON} we may ignore the presence of T_2 and consider that
it is the smaller channel resistance of T_1 which controls the discharging time constant.

Exact analysis is complicated by the fact that constant transistor
currents are not present during the transition periods, and hence
drain-to-source channel resistances are not constant. The detailed
analyses may be found elsewhere [7, 9, 16], but if we assume that
V_{TH} for the depletion loadmost transistor T_2 is $-0.8\ V_{DD}$ and V_{TH} for
the enhancement switching transistor T_1 is $0.2\ V_{DD}$, see Eq. (2.13),
then the equations may be simplified to:

$$t_{ON} \simeq 4\ \frac{C_{OUT}}{\beta_{enh}\ V_{DD}}$$

and   (2.15)

$$t_{OFF} \simeq 4\ \frac{C_{OUT}}{\beta_{dep}\ V_{DD}}$$

where β_{enh} and β_{dep} are the transistor gain factors [see Eq. (2.6)]
for the enhancement and depletion mode transistors, respectively.

Thus, the switching times are inversely proportional to the transistor gain factors. However, since this parameter is itself a function of
the transistor aspect ratio W/L, the above equations may be further
expressed in terms of W/L. With final rounding up to more exactly
match measured results, we have the rule-of-thumb values:

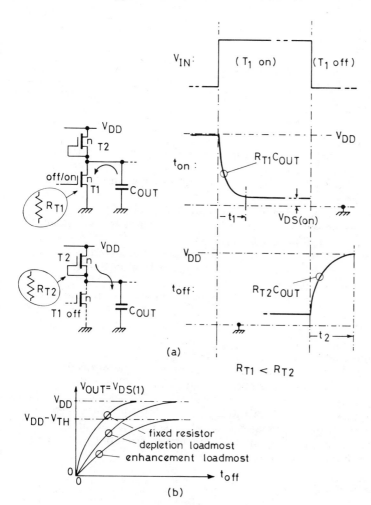

(a)

$R_{T1} < R_{T2}$

(b)

FIGURE 2.22 The switching speed of ratioed nMOS: (a) the simple INVERTER gate with depletion loadmost T_2; (b) t_{OFF} for fixed resistor load R_D, depletion loadmost, and enhancement loadmost.

$$t_{ON} \simeq 30 \left(\frac{L}{W}\right)_{enh} C_{OUT}$$

and

$$(2.16)$$

$$t_{OFF} \simeq 60 \left(\frac{L}{W}\right)_{dep} C_{OUT}$$

where t is in nanoseconds and C in pF. These final results are generally in the form one would expect, since the smaller the aspect ratio of a transistor (W/L small, L/W high) the higher its channel resistance, and the longer the time-constant associated with it. It should be mentioned that t_{OFF} of a loadmost circuit cannot be as fast as that provided by a fixed-value resistor R_D, where R_D is equal to the loadmost channel resistance in the logic 0 output state, but on the other hand is greatly superior to that provided by an enhancement transistor as initially shown in Figure 2.20c. This comparison is indicated in Figure 2.22b. Lastly, because the aspect ratio W/L is the most significant single parameter in determining the switching times, design engineers at the silicon level may often employ a normalized timing monograph for a given fabrication process, incorporating W/L = 1 and C_{OUT} = 1 pF. From such a normalized graph of T_{OFF} versus output voltage swing for various values of V_{DD}, actual T_{OFF} for C_{OUT} = x and W/L = y may be obtained, being x/y times the value read off the normalized graph. T_{ON} may likewise be determined, or taken as a certain fraction of the T_{OFF} value.

Finally, the power taken from the dc supply rail by a MOS logic gate may be appreciated as being effectively zero in the static logic 1 output state, assuming no current is being drawn from the gate output, but in the static logic 0 output state will be given by

$$V_{DD} \frac{V_{DD}}{R_D + R_{ON}} \text{ mW} \qquad\qquad (2.17)$$

if V is in volts and R in kΩ. Since $R_D \gg R_{ON}$, this may be simplified without appreciable error to

$$\frac{V_{DD}^2}{R_D} \text{ mW} \qquad\qquad (2.18)$$

$$= 0.5 \frac{V_{DD}^2}{R_D} \text{ mW}$$

assuming a 50% on/off duty cycle for the gate. Typically the logic 0 static power dissipation per gate may be, for example, $\frac{1}{2}$ mW or less depending upon device geometry, but even though this is a small value it nevertheless imposes an upper limit upon the number of gates per chip which may be present, without exceeding the maximum chip dissipation rating.

Turning to CMOS logic circuits, we have two immediate advantages over single-channel MOS circuits, namely:

1. Ideally zero gate power dissipation in both the static logic 0 and the static logic 1 output state
2. Low and ideally equally output impedances in both the logic 0 and logic 1 states

The former means that more gates per chip can be considered without running into chip dissipation problems, while the latter materially increases the t_{OFF} switching speed in comparison with our previous considerations. The disadvantage is a somewhat more complex fabrication process.

The simple CMOS INVERTER logic gate is shown in Figure 2.23a. Both the n-channel and the p-channel devices are now enhancement-mode, being off when $V_{GS} = 0$. For the n-channel device T_1, V_{GS} has to be made positive to cause T_1 to conduct, while for the p-channel device V_{GS} has to be made negative to cause it to conduct. The action of this circuit is entirely straightforward; with V_{IN} at logic 0 (below V_{TH} of T_1) T_1 will be off, but T_2 will have a forward gate bias of $|V_{DD}| - |V_{IN}|$, $\simeq |V_{DD}|$, and hence is fully on; conversely, when V_{IN} is at logic 1 (high), T_1 will be forward-biased, but $V_{GS(2)}$ will now be below its threshold value, and T_2 will be "off." Thus the two circuit states are:

T_1 fully on, T_2 off, output = logic 0

T_2 fully on, T_1 off, output = logic 1

The logic 0 and logic 1 output voltages will be almost identically 0 V and V_{DD}, respectively, the output impedance in the 0 state being the "on" channel resistance of T_1, the output impedance in the 1 state being the "on" channel resistance of T_2.

It will further be appreciated that there is no necessity for the aspect ratios of the p- and n-channel transistors to be ratioed in the manner that was necessary in single-channel logic circuits. CMOS circuits are therefore sometimes said to be "ratioless" circuits. However, in the detailed silicon layout the designer still has the ability to adjust the W/L aspect ratios for optimum circuit performance, such as equal output impedances and switching speeds, bearing in mind that the

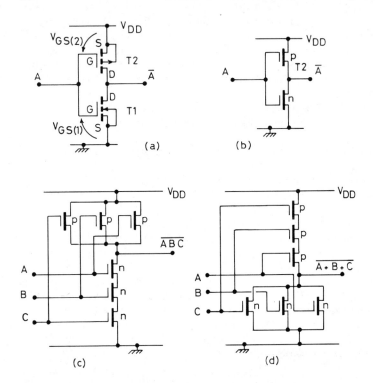

FIGURE 2.23 Ratioless CMOS logic circuits: (a) simple INVERTER
gate, precise symbols; (b) usual simplified format; (c) 3-input NAND
gate; (d) 3-input NOR gate.

hole carrier mobility in the p-channel devices is some $2\frac{1}{2}$ times slower
than the electron carrier mobility in the n-channel devices. To account
for this physical factor, the W/L ratio of the p- channel devices is
usually made about twice the W/L ratio of the n-channel devices.

Figure 2.23c and d illustrate CMOS NAND and NOR gates. The
complementary circuit configuration of the upper p-channel and the
lower n-channel devices will be apparent. Complex series/parallel
logic configurations are equally and readily possible.

The transfer characteristic of the CMOS INVERTER is given in Fig-
ure 2.24. Switching commences at $|V_{TH}|$ away from the rail voltages.
However, since $|V_{TH}|$ is largely proportional to V_{DS}, the range of
supply voltage variations and noise levels which can be accommodated
is large. NAND, NOR, and other logic configurations have similar
transfer characteristics, but may show minor hysteresis depending
upon the pattern of energization of the several inputs [8].

The static power per gate is extremely small, being V_{DD} × the small leakage current of the off transistor. Since the latter may be nA, this gives a static power dissipation of only nW per gate. During transitions between logic 0 and 1 and vice versa, however, there will be transitory rail-to-rail current flow, and also charging and discharging of the output capacitance C_{OUT}. During the charging of C_{OUT}, $\frac{1}{2}C_{OUT}(dV)^2$ joules of energy is stored in the capacitor, where dV is the change of voltage across the capacitor, with a similar amount of energy being dissipated in the source resistance from V_{DD}; during the discharge of C_{OUT}, this stored energy is dissipated in the sink resistance to 0 V. Hence, if the gate is operating at f_g operations per second, the power taken from the supply due to the presence of C_{OUT} is $C_{OUT}(dV)^2 f_g$ watts, which in most cases becomes

$$C_{OUT} V_{DD}{}^2 f_g \quad W \tag{2.19}$$

It may be noted the transition time for which there is a direct rail-to-rail current flow is generally much shorter than the time taken to fully charge or discharge C_{OUT}, and hence the power component of the rail-to-rail current is normally negligible compared with the capacitor power consumption feffect. The total chip dissipation, therefore, is a function of the number of gates and their operations per second, which may impose an upper limit on the allowable number of gates per chip.

Finally, the CMOS logic transmission gate (or "pass gate") must be noted. Such a circuit is completely impractical in bipolar technology

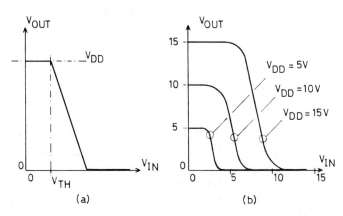

FIGURE 2.24 The transfer characteristics of the CMOS INVERTER: (a) basic characteristic; (b) family for varying supply voltage V_{DD}.

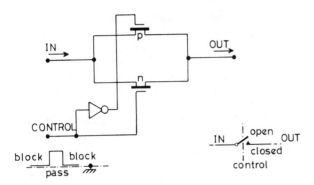

FIGURE 2.25   The simple CMOS logic transmission gate. (Note: input and output are interchangeable, allowing two-way transmission.)

due to the base electrode not being insulated from the collector and emitter electrodes, but is readily available in insulated-gate CMOS technology. It restores to logic designers the flexibility which was available to previous-generation switching circuit designers using relay contacts and similar circuits [31]. The basic transmission gate is shown in Figure 2.25. With the control input at 0 V, neither the n- nor the p-channel device will allow the transmission of a logic input voltage to the output; with the control input at logic 1, then the p-channel device will allow the transmission of a logic 1 (high) input signal to the output, while the n-channel device will allow the transmission of a logic 0 (low) input signal to the output. In concert, therefore, the two transistors pass either logic signal from the input to the output, giving us the equivalent of a transmission switch. Note that with this sample circuit, both the control and the input voltages require to be within normal logic 0 and logic 1 voltage levels; for the transmission of analog signals which may be ± within certain bounds, then a net of two pairs of transistors will be necessary.

2.3.3   Final Circuit Comments

Further details of single-channel and CMOS logic circuits and their use may be found in many general texts [32–34]. However, before we continue with a review of basic bipolar and MOS fabrication, let us briefly note a few final points in connection with MOS circuits, in view of the increasing applicability of MOS to the whole custom and semicustom fields.

Firstly, "scaling" of MOS technology. We have seen in the previous developments how the channel width W and the channel length L play an important part in MOS and CMOS circuit design. Equally it is well appreciated that the minimum geometry sizes of all integrated circuits are still decreasing, both in the bipolar and the MOS area. However, the scaling down of MOS geometry has very significant effects if certain other MOS parameters are likewise scaled.

Let us consider the reduction of L and W by a linear factor α, the scaling factor, where $\alpha > 1$. In order to maintain the same electric field and other physical parameters within the transistor, so that the device is operating under substantially the same internal conditions, in total the following factors require rescaling:

Factor	Modification
Channel length L	$1/\alpha$
Channel width W	$1/\alpha$
Gate oxide thickness	$1/\alpha$
Supply voltage	$1/\alpha$
Current	$1/\alpha$
Doping concentration	α
Channel resistance	Unchanged

The results of this scaling are:

Factor	Modification
Gates per unit area	α^2
Power per gate	$1/\alpha^2$
Gate capacitance C_{ox}	$1/\alpha$
CR time constants and resultant gate delays	$1/\alpha$
Power/delay product	$1/\alpha^3$

Note the improvement in gate delay arises from the reduction in circuit capacitance, resistance values remaining approximately constant.

Hence, scaling of MOS logic circuits appears to be a very powerful means of increasing not only the maximum number of gates per chip, but also the circuit performance. Disadvantages include:

Factor	Modification
Resistance of local metalization interconnect $\rho \dfrac{\ell}{A}$	α
Resistance of across-chip metalization interconnect $\rho \dfrac{\ell}{A}$, assuming chip size and hence ℓ unchanged	α^2
Current density in metalization interconnect	α

Thus, although gate speeds may be increased by scaling, this increase may be difficult to propagate across the total chip interconnect floor plan. Finally, scaling below 2 μm geometry becomes increasingly limited by physical features, so that there is a gradual limit to what may be achieved by direct scaling techniques. Nevertheless, the technique is a supremely powerful advantage of MOS and CMOS technology, one which is not present to such a useful degree in bipolar technology. For further details see various references [9, 16, 35–40].

   Next, comment should be made concerning the polarity of the substrate in MOS and CMOS circuits. Earlier we noted that the substrate should be at the most negative potential of the circuit for n-channel devices (most positive for p channel), and in general we have assumed it to be electrically connected to the source terminal (see Figs. 2.16c and 2.23a). From our earlier considerations and in particular Figure 2.18, it will be apparent that the substrate voltage, if different from that of the source electrode voltage, will modify the V_{GS} cutoff voltage; indeed even in normal operation there are voltage gradients within the substrate (the "body effect") which need to be considered when evaluating the device characteristics in detail.

   If the substrate is given a specific biasing voltage V_{BS} with respect to the source terminal, then reverse bias (V_{BS} negative for n-channel, V_{BS} positive for p-channel devices) will result in an increased value of $|V_{TH}|$ before drain-to-source conduction commences. Note that such "back-biasing" always increases the magnitude of the threshold voltage V_{TH}, making V_{TH} more positive for p-substrate n-channel devices and more negative for n-substrate p-channel devices (see Fig. 2.17); it cannot be used to decrease the magnitude of V_{TH}. While increase of $|V_{TH}|$ is not a primary advantage, the presence of a reverse-biased substrate has certain alternative advantages, such as reducing junction capacitances and the complex voltage gradients otherwise present within the substrate. Thus, an increase in speed-power product is possible.

The requirement to provide a separate substrate bias voltage supply means that it is not a technique usually adopted in standard logic gates. However with regular structures, in particular random-access memory/read-only memory (RAM/ROM) circuits, such a technique may be present; the required substrate voltage may be a separate dc supply to the chip, or generated by diode rectification of the output from an on-chip substrate bias oscillator circuit. Further details may be found elsewhere [7, 21], but we shall have little need to consider this feature again.

Finally, mention must be made of analog circuits. While custom analog circuits do not currently have the glamour and publicity of custom digital circuits, they still represent a very wide range of potential applications, one of possibly increasing significance in custom-specific microelectronics.

The problem of applying MOS and CMOS to analog applications has largely been that of achieving linearity of amplification and temperature stability from the MOS transistors. In addition, further problems, such as low g_m, stability of performance with process variations, and difficulties of matching input and output voltage levels in cascaded stages have been prominent. Finally, the presence of high gate-to-drain capacitance C_{gd} due to the misalignment and overlap of gate and drain electrodes, giving rise to a high Miller effect in common-source amplification mode, was initially prevalent.

Many of these early problems have now been overcome. In particular:

1. CMOS technology provides ready means of adjusting quiescent voltage levels within the amplifier, thus facilitating cascading of stages.
2. The superceding of metal-gate by silicon-gate construction (see following sections) stabilizes gate threshold values.
3. Self-aligning gates (to be discussed later) minimizes C_{gd}, and hence reduces the Miller effect and its resulting high-frequency circuit degradation.

Nevertheless, the matching performance of MOS transistor-pairs and $1/f$ noise still tends to be inferior to that available from bipolar technology.

Figure 2.26 shows circuit details of a particular CMOS analog circuit; significant is the use of current mirrors formed by two p-type transistors below the V_{DD} rail. Further details of ongoing analog developments may be found elsewhere [41–46]. The specific LinCMOS technology of Texas Instruments, which claims analog performance equal to bipolar [47], is indicative of current developments in this area, which will reinforce the significance of CMOS compared

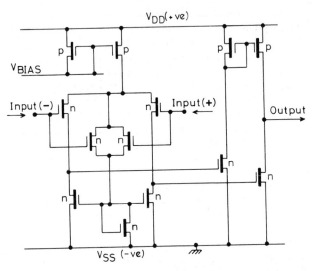

FIGURE 2.26   A CMOS array linear comparitor amplifier. Gain-band-
width product 5 MHz, small-signal voltage gain 90 dB, common-mode
input voltage range 0 to $(V_{DD} - 1.5)$ V, input offset voltage 50 mV,
output slew rate 100 V/μs. (Details courtesy Intersil, London [46].)

with other microelectronic technologies over a wide spectrum of potential applications.

2.4 *n*MOS AND *p*MOS FABRICATION

Having covered this broad survey of MOS device and circuit developments, we may now continue with the various fabrication structures of
MOS technology. Like the bipolar structures covered in Sec. 2.2, we
will show basic concepts so that the reader may be generally aware of
the principles and practice involved.

Figures 2.16 and 2.17 gave the general cross-sectional structure
of *n*- and *p*-channel metal-gate devices. In Figure 2.27 we have the
more modern structure of the buried silicon-gate device, which shows
a considerable performance improvement over metal-gate construction,
due largely to the more accurate control over the fabrication of the
gate.

The basic steps in the fabrication of the silicon gate are shown in
Figure 2.28. Two important points may be noted:

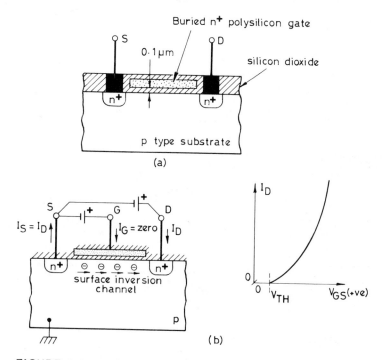

FIGURE 2.27   The buried silicon-gate n-channel enhancement-mode FET: (a) general nMOS construction with buried n^+ polysilicon gate (pMOS construction basically similar); (b) action with V_{GS} positive, as in Figure 2.17a).

1.   The polysilicon gate and other buried polysilicon connections are grown at an early fabrication stage, and are not influenced by further fabrication steps.
2.   The cut for the eventual source and drain electrodes is made across the already formed gate electrode (see Figure 2.28e), the n^+ implantations to form the source and drain being spaced apart by the presence of the actual gate.

The great importance of the latter feature is that apart from a very minor diffusion under the edges of the gate (undercut), considerably exaggerated in Figure 2.28f, the gate electrode does not overlap the source-and-drain electrodes. This "self-aligning" process eliminates the overlap which is inherent in metal-gate fabrication, thus reducing the Miller effect and allowing smaller device geometries to be reliably

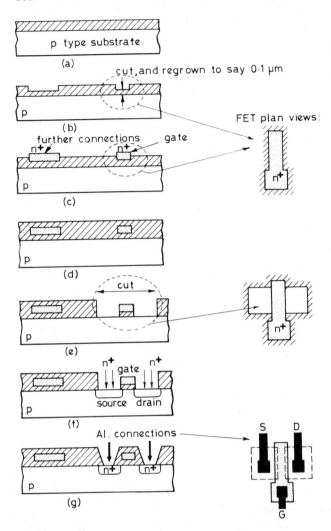

FIGURE 2.28   The self-aligning process with *n*MOS silicon-gate construction (idealized and not to scale; *p*MOS silicon-gate similar):
(a) substrate and oxide; (b) definition for buried connection layers;
(c) formation of low-resistance n^+ polysilicon gates and other connections; (d) oxide regrowth; (e) cut through oxide for source and
drain diffusions; (f) n^+ diffusions for source and drain; (g) oxide
regrowth and final via cuts to source and drain for final connections.

realized, both of which provide the increased performance capability of silicon-gate devices compared with metal-gate. A disadvantage, however, is that the gate breakdown voltage is lower, and metal-gate construction may still be appropriate where on-chip voltages of, say, 10 V or more are required.

The critical dimensions of the silicon-gate construction are shown in Figure 2.29. W and L are the gate width and length, respectively, L_{DIFF} is the lateral diffusion under the gate edges, T_{OX} is the oxide thickness, X_J is the junction depth, and D_S is the substrate doping level. Correct scaling of the MOS device, as discussed in the previous section, requires all these physical parameters to be identically scaled. Clearly there are eventual physical limits to the degree by which a structure can be scaled down.

The p-channel devices have similar structures to that illustrated in Figures 2.27 and 2.28, except that an n-type substrate is employed and source and drain implants are p^+ rather than n^+.

Certain other nMOS structures may also be found which have specific advantages but also disadvantages. Two are illustrated in Figure 2.30. The double-diffused metal-gate (DMOS) structure is shown in Figure 2.30a; this has the advantage of providing an extremely short channel length, not dependent upon photolithography limitations, being the width of the p^+ channel under the gate electrode. The π region surrounding the drain terminal is almost intrinsic silicon, and provides a drift channel to complete the source-to-drain conduction path. Very high speed operation, with gate delays of 1–2 ns, and

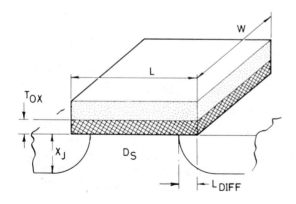

FIGURE 2.29 Cross-section of a silicon-gate device, giving the critical scaling variables.

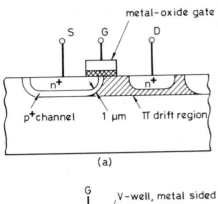

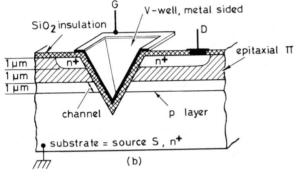

FIGURE 2.30   Further *n*-channel fabrications: (a) double-diffused MOS (DMOS); (b) V-groove MOS (VMOS).

high voltage operation are claimed for DMOS devices, but at the expense of additional mask and processing stages [13].

Figure 2.30b shows the V-groove VMOS structure. Again, very short channel lengths and potentially high voltage capability are objectives of this construction, but at the expense of a much more difficult fabrication philosophy. It will be noted that there is an appreciable channel border area around the V groove, thus increasing the W/L ratio, and hence high source to drain current levels are possible; such devices may be more relevant for power applications than for the low current, low power purposes usually associated with MOS technology [8]. The precise shape of the V groove is dictated by the crystal structure of the silicon substrate, which can be etched at an angle of 55° into the silicon along one of the crystal planes of symmetry [49]. Further details of VMOS technology, including the always-conducting depletion-mode FET action which is present between the π region and the drain, may be found elsewhere [7, 8, 50]. It is possibly true to say that VMOS still has fabrication and yield problems, and may, therefore, find its main use in high-power switching applications and the like rather than in large-scale (LSI) or very large-scale integration (VLSI) situations.

Further details of both metal-gate and silicon-gate structures may be found in more detailed publications [24-30, 40, 48].

2.5 CMOS FABRICATION

Turning to CMOS fabrication, which currently represents the most dynamic area, we find that the basic structures which we have already covered will be retained, but additional process steps and device isolation are necessary in order to combine both n-channel and p-channel devices on the one substrate.

Three basic CMOS structures are illustrated in Figure 2.31. The first is an n-type substrate, into which a p-type deep well (p tub) has been implanted to form the "substrate" of the nMOS transistor. The second employs a p-type substrate, with a n-type deep well (n tub) for the pMOS transistor. A further variant (not shown) is a twin-tub structure, in which both a p- and a n-type deep well are implanted into a substrate for the n-channel and p-channel transistor, respectively. Both Figures 2.31a and b are metal-gate construction CMOS. The equivalent buried silicon-gate structures, such as shown in Figure 2.31c, represent more recent practice, where the silicon gates are each self-aligning, as covered in the previous section. Indeed all the principal features previously considered carry over to the CMOS area.

Isolation requirements between devices in CMOS structures are critical. One of the early problems with CMOS structures using silicon substrates was the danger of latch-up through the formation of parasitic thyristors, which if present and triggered "on" by high voltage or circuit impulses gave rise to excessive rail-to-rail current. Consider the four adjacent complementary doped regions shown in Figure 2.32, which will be appreciated as present in CMOS structures. Such a structure is basically a thyristor, and if the large-signal current gains $h_{FB(1)}$ and $h_{FB(2)}$ of the two transistors become such that $\{h_{FB(1)} + h_{FB(2)}\} > 1.0$, then the circuit will latch "on" in a normal thyristor manner when triggered by some initial current impulse. For further details of $pnpn$ thyristor action, see any standard text [22, 23]. This action in CMOS circuits is sometimes referred to as a parasitic "hook" action [42]; once triggered on, there is no way of switching off the thyristor save that of reducing the current below the device holding-current level, which effectively means the disconnection of the supply voltage in this case.

An early method of preventing latch up and providing efficient device isolation on the silicon substrate is shown in Figure 2.33. In

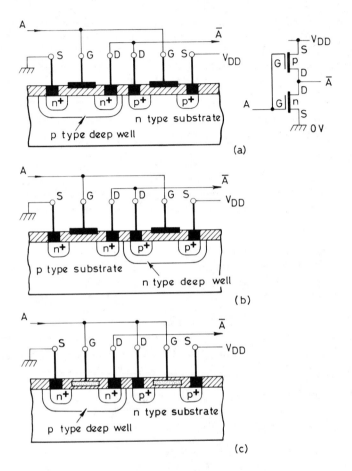

FIGURE 2.31   The basic complementary MOS (CMOS or COSMOS) construction: (a) metal-gate, p-type deep well; (b) metal-gate, n-type deep well; (c) silicon-gate, p-type deep well.

this p-tub metal-gate CMOS structure, a separate heavily doped guardband channel-stop (moat) is diffused around each n-channel and each p-channel device to provide device isolation. The n^+-channel stop between the devices prevents the formation of a parasitic transistor action between the source region of the p-channel transistor and the p well of the n-channel transistor; the p^+-channel stop then prevents the formation of parasitic action between the drain region of the n-channel transistor and the n^+-channel stop.

This dual guardband isolation with metal-gate devices represents first-generation CMOS design practice. The technique may be found

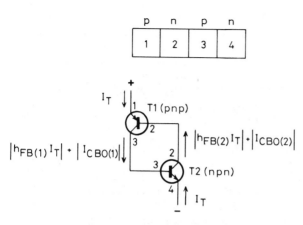

FIGURE 2.32   The parasitic thyristor action possible with adjacent
p, n, p, n areas.

in the 4000 series SSI/MSI off-the-shelf CMOS packages, but the area
required per transistor pair is large and hence not applicable to LSI
and VLSI applications. Its advantage is that high breakdown voltages
can be provided, allowing working voltages of 18—30 V if required,
with latch-up immunity to overvoltage and circuit spikes.
    Replacement of the dual guardband by a single guardband, still
with metal-gate transistors, represented the first attempt to modify
device size and obtain better packing density, and the beginning of
possible LSI and semicustom applicability. In general, the very high

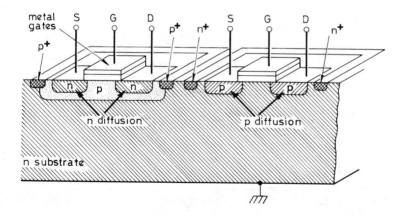

FIGURE 2.33   Dual guardband channel stops used for device isolation
in first-generation CMOS (not to scale).

voltages available with the dual guardband were not available with the single guardband. However, considerable improvements both in performance and packing density resulted with the development of the silicon-gate combined with oxide device isolation techniques. This is representative of all modern CMOS developments, and has opened the way to high-speed LSI, VLSI, and custom-specific applications. The penalty to be paid for these improvements is generally lower operating and breakdown voltages, about 8 V maximum.

Oxide isolation and self-aligned silicon gate technology is represented by the LOCMOS (local-oxide complementary MOS) technology of Philips, Eindhoven (The Netherlands), and the IsoCMOS (isoplanar CMOS) technology of Mitel, Canada, and others. Figure 2.34 shows the general structure of the IsoCMOS process; not unlike the oxide isolation technique described in Sec. 2.2.4 for the isoplanar I^2L technology, it provides efficient isolation between devices, using less silicon area than p^+ or n^+ guardbands, and allowing device areas to be

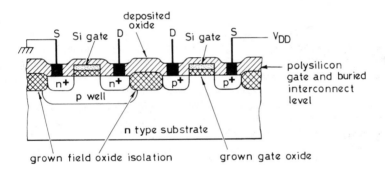

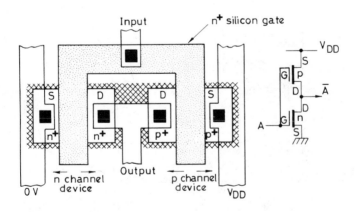

FIGURE 2.34   The high-performance oxide-isolated separate p-well IsoCMOS structure (not to scale).

shrunk by permitting the source and drain diffusion areas to abut the field oxide isolation.

The principal steps in the fabrication are as follows, omitting certain stages which may be present to adjust V_{TH} and other detailed device parameters:

1. Grow thin oxide layer on n-type substrate (SiO_2)
2. Deposit layer of silicon nitride (Si_3N_4)
3. Cover with photoresist
4. Apply mask no. 1 to define the areas of all the final p- and n-channel transistors
5. Cut through to substrate leaving all transistor areas as Si_3N_4/SiO_2 islands
6. Re-cover with photoresist
7. Apply mask no. 2 to define the deep p well carrying each n-channel transistor
8. Cut through photoresist; perform a high-concentration boron implant and drive in to form the complete deep p well
9. Remove all photoresist, and grow *thick* field oxide between all Si_3N_4/SiO_2 islands (= final device isolation)
10. Remove Si_3N_4/SiO_2 islands
11. Regrow thin gate oxide (SiO_2) over exposed substrate surface
12. Deposit polysilicon layer across whole surface
13. Diffuse phosphorous dopant into polysilicon to create the n^+ gate and buried interconnection level (poly level)
14. Apply mask no. 3 to define the polysilicon gate and other interconnections
15. Etch away all unwanted polysilicon leaving thin oxide
16. Apply mask no. 4 to define the overall n-channel transistor areas
17. Cut through thin oxide to form n-channel source and drain electrodes, and diffuse in n^+-type areas (NOTE: the pre-defined polysilicon gate provides the self-aligning feature)
18. Apply mask no. 5 to define the overall p-channel transistor areas
19. Cut through thin oxide to form the p-channel source and drain areas, and diffuse in p^+ type areas
20. Deposit oxide across whole surface
21. Cover with photoresist
22. Apply mask no. 6 to define all contact windows (vias) to the source, drain, and polysilicon interconnect points
23. Etch all contact windows
24. Deposit aluminum over whole surface area
25. Cover with photoresist
26. Apply mask no. 7 to define the required metalization interconnect pattern

27. Etch away all unwanted metal
28. Deposit final protective passivation (Vapox) over whole sur-
     face area

It may be noted that the thick field oxide which forms the device isolation and the thin gate oxide under the polysilicon gates are grown
on the silicon substrate. The later layers have to be deposited. There
may also be other process steps between those listed above, or other
variations, but the fabrication steps for oxide-isolated CMOS will generally be as above. Further details may be found published [10, 39,
42, 51-53].

The overall effect of oxide-isolation applied to silicon-gate CMOS
technology has been to increase both the speed and the packing density by a factor of three or more in comparison with metal-gate CMOS.
The only comparable bipolar technology for LSI and VLSI applications
is I^2L and its variants, but as will be surveyed in the following section
there now appear to be noticeable advantages in CMOS over I^2L.

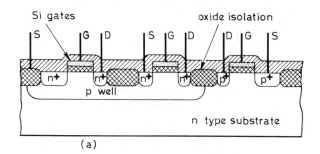

(a)

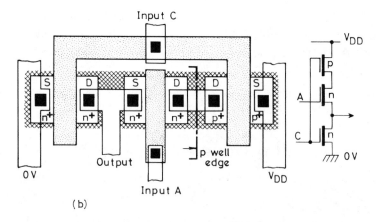

(b)

FIGURE 2.35   The "ubiquitous p well" (or common p well) oxide-
isolated CMOS (not to scale). (a) General structure; (b) plan.

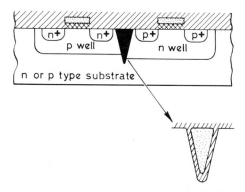

FIGURE 2.36   The twin-tub CMOS construction, with deep-groove
isolation between p- and n-channel devices.

There are, however, further variants in CMOS fabrication which
we will briefly note. Figure 2.35 illustrates the "ubiquitous p-well"
approach of American Microsystems in particular, which employs a
common p well in an n-type substrate for several n-channel transistors
[41]. Grown field-oxide isolation is retained for isolation between the
p well and adjacent p-channel transistors, so as to prevent any parasitic p-channel conduction between these two p regions. The particular advantage of this approach is in memory circuit applications,
where more n-channel devices than p-channel devices may be used;
this is in contrast to the balanced use of p- and n-channel devices in
usual CMOS logic circuits.

An alternative construction is a twin-tub structure, using either
an n- or a p-type substrate, with a deep p well for the n-channel
transistors and a deep n well for the p-channel transistors. Field
oxide isolation similar to that shown in Figure 2.34, or deep-groove
isolation as shown in Figure 2.36 may be found. The case for twin-tub
structures is that they allow greater flexibility for device optimization
without any increase in silicon area, and hence this may be a fabrication technology of increasing importance [39, 54].

It is also possible to consider the fabrication of CMOS circuits on
some form of insulated substrate other then the bulk-silicon p- or
n-type substrates which have so far been considered. Figure 2.37
illustrates one proposed structure wherein the MOS transistors are
built up as islands on the surface of the silicon dioxide insulating layer
[39]. Since there are no bulk silicon substrate effects now present, it
is claimed that closer control of the p- and n-channel device characteristics may be possible.

More widely pursued, particularly for military and space applications where high radiation tolerance is required, is the insulated substrate approach shown in Figure 2.38. The substrate here is sapphire

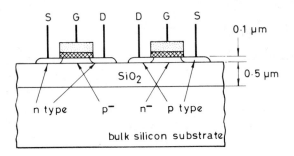

FIGURE 2.37   Silicon-on-oxide insulated substrate CMOS (SOI-CMOS) (not to scale).

upon which n-type silicon is deposited to form the starting layer of the device fabrication. Oxide isolation is present between devices similar to the IsoCMOS fabrication. The advantages claimed for SOS-CMOS are that fewer process steps are involved and higher perform-ance due to reduced size and circuit capacitances is possible; an increase of speed and packing density of two or more over conventional silicon-gate CMOS processes has been claimed. The disadvantages are

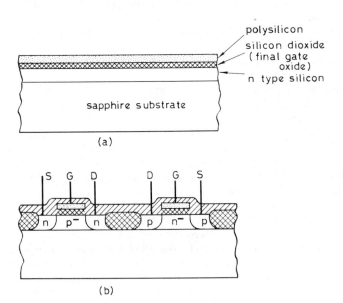

FIGURE 2.38   Silicon-on-sapphire insulated substrate CMOS (SOS-CMOS) (not to scale). (a) Preliminary structure before selective maskings; (b) final structure.

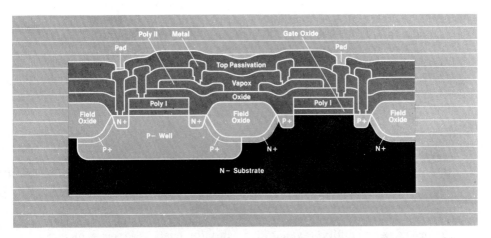

(a)

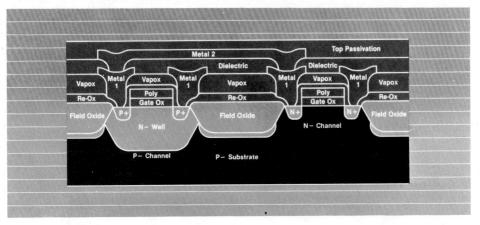

(b)

FIGURE 2.39   Multilayer interconnection CMOS constructions:
(a) dual-polysilicon P^2CMOS; (b) dual-metal M^2CMOS. (Details courtesy National Semiconductor Corporation, Santa Clara, CA.)

that the cost of the sapphire substrate is high, heat conduction through
the substrate is poor, and differential expansion between the silicon and
sapphire can be troublesome. Nevertheless it represents a basically
simple, very high-performance technology, limited at present mainly
by thermal problems [8, 10, 13, 39, 41, 55].

We have so far considered only single-layer metalization in the various idealized structures we have illustrated. The later CMOS structures admittedly gave us two layers of interconnect, the buried polysilicon level and the final metalization level. Even so, difficulty in

interconnection of a complete chip architecture may be present. To complete this general survey, therefore, Figure 2.39 shows the structure of two more complex CMOS processes, the two-level polysilicon interconnect P^2CMOS process and the two-layer metal interconnect M^2CMOS process [42]. These are possibly indicative of the increasing interconnect complexity which may grow up around the basic CMOS fabrication process.

2.6 A COMPARISON OF AVAILABLE TECHNOLOGIES

There is no one technology which is supreme above all others for all applications. Each has its particular merits or devotees, and a choice for a specific application may be made on the grounds of appropriate performance capability, availability, and /or cost. Extremes of performance are easy to define; for example, if absolute speed was supreme, then bipolar ECL technology would be appropriate, but if absolute minimum standby power was critical, then CMOS would be the choice.

Between such extremes runs a continuum of performance availability, which we will generally survey herewith. Figure 2.40 indicates the gate delay versus average power per gate for the various variants

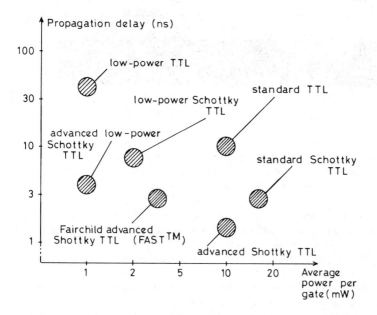

FIGURE 2.40   Power versus gate delay for various TTL families.

TABLE 2.4   Speed and Power Data for Bipolar Logic Families

Family	Typical gate delay (ns)	Typical power per gate (mW)	Power-delay product pJ	Fabrication masks required
Standard TTL	10	10	100	7
Advanced Schottky TTL	2	10	20	8–9
Advanced low-power Schottky TTL	4	1	4	8–9
Standard I^2L	15	0.05	0.8	8
I^3L	4	0.05	0.2	8–10
Integrated-Schottky logic (ISL)	4	0.4	1.6	9–10
Schottky-transistor logic (STL)	2	0.4	0.8	12
Standard ECL	1	20	20	7–10
Standard CML	8	1	8	7–8
High-speed CML	3	0.3	1	7–8

TABLE 2.5 Speed and Power Data for MOS Logic Families

Family	Typical gate delay (ns)	Typical power per gate (mW)	Power-delay product pJ	Fabrication masks required
pMOS	50	1	50	6
nMOS	15	0.5	7.5	6–8
High-voltage metal-gate CMOS	100	2.5	250	7–8
Low-voltage silicon-gate CMOS	30	0.2^a	6^a	7–8
Oxide-isolated silicon-gate CMOS	10	0.15^a	1.5^a	7–10
Silicon-gate SOS-CMOS	3	0.15^a	0.45^a	7–8
DMOS	2	0.12^a	0.24^a	7–8

[a] At 1 MHz, V_{DD} = 5 V, C_{OUT} = 5 pF; static dissipation zero.

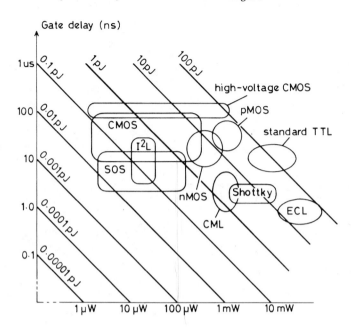

FIGURE 2.41   Power versus gate delay for several technologies.

of TTL technology.   The conflict between low power and gate delay
will be apparent.

A comparison across the wider family of bipolar technologies is
given in Table 2.4.   The supremacy of ECL for highest speed capability will be noted, but this is closely approached by the high-speed
current-mode logic, which uses reduced internal voltage excursions
in its nonsaturating operating mode.

A similar comparison across the family of pMOS, nMOS, and CMOS
technologies is given in Table 2.5.   The general supremacy of the later
CMOS variants will be clear, which supremacy should advance with
continued scaling and also further insulated substrate developments.

Should we attempt to summarize the typical gate delays and power
per gate for all families, a picture such as that given by Figure
2.41 may be compiled.   It is difficult to define the power dissipation per gate for CMOS derivatives, since this is dependent upon
$C_{OUT} V_{DD}{}^2 f_g$ [see Eq. (2.19)].

Table 2.6 gives further comparative figures across the bipolar and
MOS families.   Improvements in processing mean that scaling of I^2L
and CMOS in particular is continuing, and hence the packing densities listed herewith may become obsolescent.

Figure 2.42 complements Figure 2.41.   The quest for high speed
allied with low power dissipation and small gate area is emphasized by
these varied comparisons.

TABLE 2.6   Packing Density and Voltage Characteristics of Principal Bipolar and MOS Families

Family	Packing density, 2-input gates/mm²	Processing steps (approx)	Supply voltage (V)	Typical logic swing (V)	Typical noise margin (V)
Standard TTL	20	22	$V_{CC} = 5$	3.2	0.4
I²L	100	23	a	<1	<0.2
ECL	20	23	$V_{EE} = -5.2$	0.9	<0.2
pMOS	100	12	$V_{DD} = -15$	2/3 V_{DD}	1.5
nMOS	150	15	$V_{DD} = 5$	2/3 V_{DD}	1.0
IsoCMOS	100	22	$V_{DD} = 5^b$	V_{DD}	$\simeq$ 1/2 V_{DD}
SOS-CMOS	150	20	$V_{DD} = 5^b$	V_{DD}	$\simeq$ 1/2 V_{DD}

[a] Internally regulated on-chip voltage, $\simeq$ 1V.

[b] May have alternative supply voltages from, for example, 1.5 to 8 V.

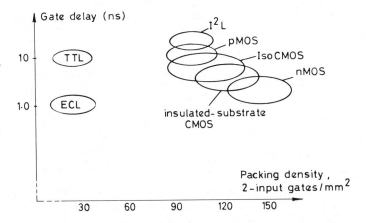

FIGURE 2.42   Packing density versus gate delay for several technologies.

Further details of device performance may be found in more detailed references [10, 13, 36-41, 48, 51-53]. Possible future developments will also be found in certain of these references, but we will return to consider this further in Chapter 6. These considerations have been confined to digital circuits, since this represents our main area of interest; for analog developments, particularly recent CMOS developments, see other more specific publications [43-47].

2.7 FABRICATION AVAILABILITY

All the technologies which we have reviewed in the previous sections are currently in production for either off-the-shelf products or custom-specific circuits, or both. The current situation is generally as follows, with bipolar technology tending towards I^2L and MOS technology tending towards CMOS for custom-specific applications:

Technology	Off-the-shelf circuits	Custom-specific circuits
TTL	Yes (All variants)	Yes (Certain variants)
ECL	Yes	Yes
I^2L	No	X

Technology	Off-the-shelf circuits	Custom-specific circuits
CML	No	Yes
nMOS	Yes	Yes
pMOS	Yes	Yes
CMOS (metal-gate)	Yes	Yes
CMOS (silicon-gate)	Yes	Yes
CMOS (insulated substrate)	No	Yes

Newer techniques which are still in the development phase, particularly gallium-arsenide technology, may be added to this list in the near future.

The 1982 production of all semiconductor devices totalled some 2×10^{10}, split approximately 60:40 between bipolar and MOS products. Standard TTL still accounted for a high proportion of the bipolar market, with memories and microprocessors dominating the MOS market. The custom-specific market, however, accounted for less than 10% of this total production value, but this is the area in which rapid growth is forecast within the next decade.

We will conclude this chapter with some additional general information on basic technology, which may come within the discussion orbit of system designers when considering custom-specific microelectronic fabrication with detailed silicon designers or fabrication vendors.

One effect of an increase in the custom-specific market will be an increased demand for mask making, to cater to the escalating number of different designs to be fabricated on standard production lines. Mask-making resources, therefore, will be a critical factor [56]. There are two types of mask in current use (1) light-field masks, and (2) dark-field masks. With the former, the majority of the mask is transparent, leaving only the significant parts of the circuit layout as opaque regions. With the latter, the significant parts are transparent, the majority of the mask being opaque.

The organic photoresist material deposited upon the wafer surface must be appropriate for the type of mask employed. The choice is between "negative" and "positive" photoresist. With the former, irradiation by ultraviolet light causes the photoresist to become chemically inert, such that the unexposed areas may be selectively removed by an appropriate chemical developer; with the latter the areas exposed to the UV radiation may be removed by the chemical developer. Between the two types of mask and the two types of photoresist we have four possibilities:

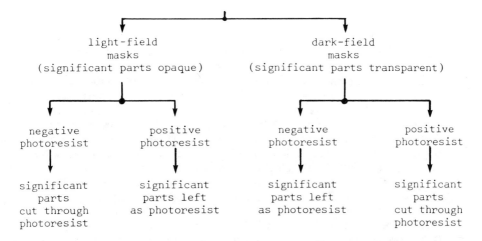

Light-field masks are preferred for ease of alignment, but dark-field masks, where the majority of incident ultraviolet radiation is blocked off from the wafer surface, may give cleaner results. One potential danger in the semicustom field is a failure of communication between mask fabrication and the production line, which results in the wrong combination of mask and photoresist. Hopefully, greater security of communication in the future will eliminate such problems.

Following each photolitographic exposure is the cutting (etching) of the appropriate pattern into the wafer surface. This may be either a wet chemical etch, a dry plasma etch, or an ion-beam milling technique. Dry plasma etch is a newer technique than the original wet chemical etch, and provides a smaller undercut than the latter, giving a dimensional limit of about 1 μm compared with the 2–3 μm limit of wet etching; however, it has yet to completely replace the original technique. Ion-beam milling is a newer and potentially much more attractive method, because it avoids all the problems with chemicals and their residues and gives dimensional accuracies of under 1 μm. It is currently capital expensive and not yet in wide-scale use.

Finally, having cut the appropriate pattern into the wafer surface, the diffusion of donor or acceptor impurities, or ion implantation or other process is made. These are complex processes, involving, for example, deposition of an impurity layer on the wafer surface and subsequent drive-in diffusion under carefully controlled temperature and timing. This is the province of the device physicist and fabrication engineer, and lies outside our detailed interests. For further information see Appendix B, or the many specialized references [7, 13, 16, 25-28, 57].

2.8 SUMMARY

In this chapter we have attempted to cover the basic principles and practice of microelectronic device fabrication and circuit design in sufficient depth for the original equipment engineer and system designer who may be involved with custom and semicustom microelectronics. In the following chapters we will continue with the principal means whereby custom-specific microelectronics are made available to the OEM designer. We will not have occasion to refer again in any detail to the underlying fabrication techniques; we will assume these to be based upon established technologies and practice already reviewed and freely available. Indeed the future expansion of custom-specific microelectronics will undoubtedly see the establishment of more small-company mask making and IC fabrication resources, such that OEM reliance upon present-day giant IC companies will be reduced. Ideally, both mask making and fabrication should be under one roof, so that interface communication between the two is optimized, and final customer-vendor interfaces are well defined.

REFERENCES

1. Fletcher, W. I. *An Engineering Approach to Digital Design*. Prentice-Hall, Englewood Cliffs, NJ, 1980.
2. Blackeslee, T. R. *Digital Design with Standard MSI and LSI*. Wiley Interscience, New York, 1975.
3. Muroga, S. *Logic Design and Switching Theory*. Wiley-Interscience, New York, 1979.
4. Mano, M. M. *Digital and Microprocessor Engineering*. Prentice-Hall, Englewood Cliffs, NJ, 1979.
5. Cahill, S. J. *Digital and Microprocessor Engineering*. Ellis-Horwood, Chichester, UK, and John Wiley, New York, 1982.
6. Lewin, D. *Logical Design of Switching Circuits*. Van Nostrand Reinhold, London, 1982.
7. Howes, M. J., and Morgan, D. V. (Eds.), *Large Scale Integrations: Devices, Circuits and Systems*. Wiley-Interscience, London, 1981.
8. Muroga, S. *VLSI System Design*. Wiley-Interscience, New York, 1982.
9. Mavor, J., Jack, M. A., and Deyner, P. B. *Introduction to VLSI Design*. Addison-Wesley, London, 1983.
10. Capece, R. P. Special report: new lsi processes. *Electronics*, 52(19):109-115, 1979.
11. Glaser, A. B., and Subak-Sharpe, G. E. *Integrated Circuit Engineering*. Addison-Wesley, Reading, MA, 1977.

12. Harris Semiconductors. Brochure *Bipolar Gate Arrays*. Harris Corp., Melbourne, FL, 1981.
13. Integrated Circuit Engineering Corp. *Basic Technology: An Overview of the Fabrication of Integrated Circuits*. Publication No. 10708, Integrated Circuit Engineering Corporation, Scottsdale, Arizona, 1980.
14. Warner, M. R., and Fordemwalt, J. N. (Eds.). *Integrated Circuits, Design Principles and Fabrication*. McGraw-Hill, New York, 1965.
15. Milnes, A. G. *Semiconductor Physics and Integrated Electronics*. Van Nostrand Reinhold, New York, 1980.
16. Sze, S. M. (Ed.). *VLSI Technology*. McGraw-Hill, New York, 1983.
17. Hodges, D. A., and Jackson, H. G. *Analysis and Design of Digital Integrated Circuits*. McGraw-Hill, New York, 1983.
18. Pierret, R. F. *Semiconductor Fundamentals*. Modular Series on Solid State Devices, Vol. I. Addison-Wesley, Reading, MA, 1983.
19. Neudeck, G. W. *The pn Junction*. Modular Series on Solid State Devices, Vol. II. Addison-Wesley, Reading, MA, 1983.
20. Neudeck, G. W. *The Bipolar Junction Transistor*. Modular Series on Solid State Devices, Vol. III. Addison-Wesley, Reading, MA, 1983.
21. Pierret, R. F. *Field-Effect Devices*, Modular Series on Solid State Devices, Vol. IV. Addison-Wesley, Reading, Massachusetts, 1983.
22. Mannone, P. Careful design methods prevent CMOS latch-up, *EDN*, 29(2):137-152, 1984.
23. Streetman, B. E. *Solid State Electronic Devices*. Prentice-Hall, Englewood Cliffs, NJ, 1980.
24. McCarthy, O. J. *MOS Device and Circuit Design*. Wiley-Interscience, New York, 1982.
25. Fairchild Corp. *Semiconductor and Integrated Circuit Fabrication Techniques*. Prentice-Hall, Englewood Cliffs, NJ, 1979.
26. Milne, A. D. *MOS Devices: Design and Manufacture*. Edinburgh University Press, Edinburgh, 1982.
27. Till, W. C., and Luxon, J. T. *Integrated Circuits: Materials, Devices, Fabrication*. Prentice-Hall, Englewood Cliffs, NJ, 1982.
28. Faggin, F., and Klein, T. Silicon gate technology. *Solid State Technol.*, 13:1125-1144, 1970.
29. Hicks, P. J. *Pre-Conference Tutorial*. Proc. 3rd Int. Conf. Semicustom ICs, London, November, 1983.
30. Altman, L. (Ed.). *Large Scale Integration*. McGraw-Hill, New York, 1976.
31. Caldwell, S. H. *Switching Circuits and Logic Design*. John Wiley, New York, 1958.
32. Sedra, A. S., and Smith, K. C. *Microelectronic Circuits*. Holt, Reinhart and Winston, New York, 1982.

33. Schilling, D. L., and Belove, C. *Electronic Circuits: Discrete and Integrated.* McGraw-Hill, New York, 1979.

34. Millman, J. *Microelectronics: Digital and Analogue Circuits and Systems.* McGraw-Hill, New York, 1979.

35. Svensson, C. VLSI physics. *Integration. VLSI J.,* 1(1):3-19, 1983.

36. Dettmer, R. The world of the DRAM. *IEE Electronics and Power,* 29:801-804, 1983.

37. Lyman, J. Scaling the barriers to VLSI's fine lines. *Electronics,* 53(12):115-126, 1980.

38. Kennedy, L. W., Smith, D. E. H., and McCaughan, D. V. Developments in LSI production technology. *GEC J. Sci. Technol.,* London, 48(2):90-96, 1982.

39. Davis, R. D. The case for CMOS. *IEEE Spectrum,* 20(10):26-32, 1983.

40. Jecmen, R. M., Hui, C. H., Ebel, A. V., Kynett, V., and Smith, R. J., HMOS II static RAMS overtake bipolar competition. *Electronics,* 52(18):124-128, 1979.

41. Wollesen, D. L. CMOS LSI: comparing second-generation approaches. *Electronics,* 52(18):116-123, 1979.

42. National Semiconductor. *The 1983 CMOS Handbook.* National Semiconductor Corp., Publication X337, Santa Clara, CA, 1983.

43. Rehman, M. A. MOS analogue integrated circuits. *Electron. Eng.,* 52(645):115-150, 52(646):75-83, 1980. 53(647):59-65, 53(648):95-111, 1981.

44. Kash, R. Building quality analogue circuits with CMOS logic arrays. *Electronics,* 54(16):109-112, 1981.

45. Kash, R. B. Analogue MOS on semi-custom integrated circuits. *PROC. IEEE WESCON,* 1980.

46. Watson, D. The realisation of analogue functions on CMOS gate arrays. *PROC. 3rd Int. Conf. on Semi-custom ICs,* paper 2/16, London, November 1983.

47. Texas Instruments. TI's new LinCMOS process launches the fast low-power linear IC future. (Advert.), *EDN,* 28(18):20-22, 1983.

48. Fairchild Semiconductors. MOS Applications. MOS Data Book, Fairchild Semiconductor Corporation, 844-858, Mountain View, CA, 1981.

49. Lee, D. B. Anisotropic etching of silicon. *J. Applied Physics,* 49:4569-4574, 1969.

50. Rogers, T. J., and Meindl, D. J. VMOS, high-speed TTL compatible MOS logic. *IEEE J. Solid-state Circuits,* SC.9:239-250, 1974.

51. Mital Semiconductor. High performance CMOS. *Electron. Prod. Des.,* 4:191-197, September 1983.

52. Richmond, P., and Broomfield, R. Upgrading system performance by IsoCMOS replacement of standard ICs. *Electronic Engineering,* 53(651):41-49, 1981.

53. Dettmer, R. What makes CMOS run. *IEE Electronics Power*, 29:547-550, 1983.

54. Payne, R. S. Twin-tub CMOS—a technology for VLSI circuits. *Tech. Dig.*, Int. Electron Devices Meeting, 1980.

55. Stebnisky, W. W., and Feller, A. State of the art CMOS/SOS technology for next-generation computers. *IEEE Proc. Int. Conf. Circuits and Computers*, pp. 329-332, 1980.

56. Penisten, G. E. A survey of the present and future status of the semi-custom arena. *J. Semicustom ICs*, 1(2):5-11, 1983.

57. Fang, F. G., and Rupprecht, R. S. High-performance MOS using the ion-implantation technique. *IEEE J. Solid-State Circuits*, SC10:205-211, 1975.

3
Programmable Devices

3.1 INTRODUCTION

The availability of programmable devices, standard volume-produced circuits which may be programmed by one or other means to provide specific circuit requirements, was discussed in Chapter 1, Section 1.4. It will be recalled from Table 1.6 that such off-the-shelf products may be either mask-programmable or field-programmable, the former normally involving the fabrication of appropriate final on-chip connections, the latter usually involving the destruction of connections already present on the chip. Mask-programmable devices are personalized to a particular customer requirement by the manufacturer or vendor connecting appropriate internal points, but field-programmable devices may be personalized by the purchaser of the device, through the application of defined electrical signals to destroy appropriate internal connecting paths.

We have seen that there were a number of categories of such devices, including:

1. Read-only memories (ROMs)
2. Programmable logic arrays (PLAs)
3. Programmable gate arrays (PGAs)
4. Programmable logic sequencers (PLSs)

In this chapter we will consider these categories in greater detail, in order to give an overall view of the range and application of these devices.

Their significance in the field of semicustom large-scale integration (LSI) will be considered in the final section of this chapter.

3.2 PROGRAMMABLE READ-ONLY MEMORIES

Read-only memories (ROMs) are combinatorial logic structures in which every possible input combination of the given number of independent binary input variables is available, and from which any output function may be synthesized by ORing an appropriate number of these terms together. It will, therefore, be appreciated that ROM is a direct hardware realization of the truth table for a given function. A simple case is shown in Figure 3.1.

The basic architecture of a simple ROM consists, firstly, of a fixed decoder assembly to generate all the 2^n minterms of the n input variables, and secondly, a programmable assembly to selectively OR together the minterms required per output function. Provision for several output functions $f_1(x)$, $f_2(x)$, . . . is invariably provided, as

INPUTS				OUTPUT FUNCTION
x_1	x_2	x_3	minterm	$f(X)$
0	0	0	m_0	0
0	0	1	m_1	1
0	1	0	m_2	1
0	1	1	m_3	0
1	0	0	m_4	0
1	0	1	m_5	0
1	1	0	m_6	0
1	1	1	m_7	1

(a)

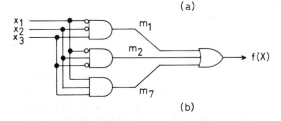

(b)

FIGURE 3.1 A simple combinatorial function. (a) Truth table, minterms m_1, m_2, and m_7 1-valued, remaining minterms 0-valued. (b) a direct realization of (a) using discrete gates.

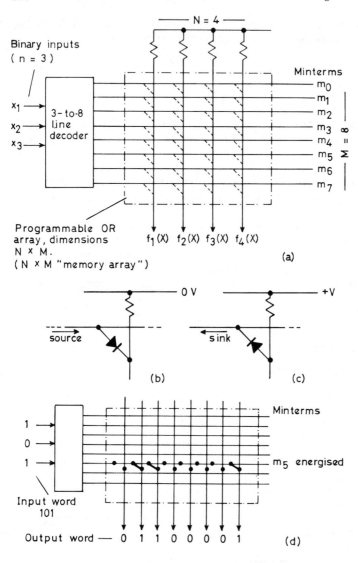

FIGURE 3.2 A basic ROM structure: (a) fixed decoder and programmable OR matrix array. (b) Diode programming links when the decoder
outputs act as sources. (c) Diode programming links when the decoder
outputs act as sinks. (d) A simple 8 × 8 ROM, showing the specific
output 01100001 generated by input 101 ("output word 01100001 stored
at location 101" in memory parlance).

shown in Figure 3.2. The major differences in different commercially available ROMs are:

1. The technology of the package, whether bipolar transistor-transistor logic (TTL) or emitter-coupled logic (ECL) or unipolar metal-oxide semiconductor (MOS) or complementary metal-oxide semiconductor (CMOS)
2. The number of inputs and outputs provided
3. The means of programming the internal OR matrix connections
4. The inhibit or enable or other facility which may be present on the individual outputs

The general hierarchy of these devices is shown in Table 3.1. Note that all are "nonvolatile" devices, that is they do not lose their internal personalization in the absence of the dc power supply, as in the case with random-access memories (RAMs) which are static or dynamic write-in/read-out bistable circuit storage arrays. Note also that the terminology ROM is usually reserved for manufacturer-(vendor-) programmed devices involving a mask and final fabrication stage for personalization, whereas programmable read-only memory (PROM) is globally applied to all other types which do not require a fabrication stage, and which may therefore be customer-(field-) programmed.

The full internal physical details of the ROM or PROM need not be intimately understood by a user. For example, among the fusible link technologies may be found polycrystalline silicon, platinum silicide, nichrome, and other materials, together with various bipolar and MOS fabrication techniques. Figure 3.3 shows details of a particular fusible structure in a TTL field-programmable range of products. In MOS technology, there is also an alternative possibility to the fusible link for field-programmable devices, this being the floating-gate technology whereby MOS transistors may be controlled "on" or "off" by a charge injected and trapped within an insulated-gate area of the MOS devices.

Further details of the internal mechanisms of ROM and other devices may be found in more detailed sources [1–7]. However, the user should consider the possible long-term reliability of PROM devices, particulary erasable PROMs (EPROMs), since in general such devices contain potential failure or degradation mechanisms over and above conventional bipolar or MOS LSI circuits [1, 7, 25–28, 41]. This must also be considered for all the further field-programmable devices covered in this chapter. Apart from such long-term considerations, the original equipment manufacturer (OEM) need only be concerned with input/output capabilities and performance.

The capacity of the ROM is usually quoted as the number of programming points within the OR matrix, for example 8 × 4 in the simple case shown in Figure 3.2a, and 8 × 8 in Figure 3.2d. Figure 3.4 shows the architecture of a 32 × 8 commercial package. For larger sizes, the

TABLE 3.1 The General Hierarchy of Read-Only Memory Nonvolatile Devices

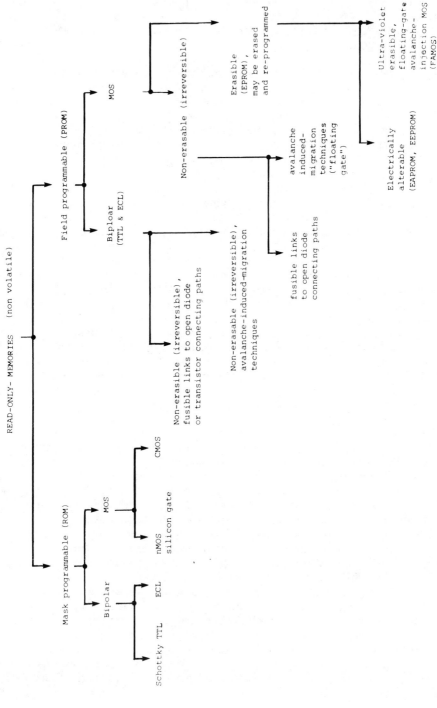

product M × N is quoted in K; for example a 4096 × 8 ROM (12 inputs, 8 outputs) would be referred to as a 32K ROM; note that a rounding off of M × N is used in the quoted K value.

For large ROM and PROM structures, a modified architecture from that shown in Figure 3.2 may be employed, in order to reduce the number of minterm output lines required from one decoder. Such an arrangement is shown in Figure 3.5b, which should be compared with the single-decode architecture of Figure 3.5a. It may be noted that the total number of programmable points in the arrangement of Figure 3.5b is $2^j \times N \times 2^{n-j}$, = $2^n \times N$, which is the same as though a single-decoder architecture was present; however, the proportions of the programmable array can now be made much more square than would be the case for a single decoder with large n, which is advantageous for good chip layout design. For very large structures, even further partitioning of the complete chip may be used; for example a 64K ROM may be partitioned into eight repeated 8K ROM structures in order to facilitate the on-chip routing of all the necessary orthogonal signal lines.

In conclusion, the ROM family of devices provides an extremely powerful means of realizing multiinput, multioutput fixed combinatorial duties, such as code convertors, binary waveform generation, speech synthesis, arithmetic functions, and the like. Advantages are that being minterm-based, no design effort is required to minimize the logic of the required input/output relationships, which is in contrast to PLAs (see Sec. 3.3), but disadvantages include:

1. Generally slower operating speeds than available from custom-designed networks, due to the internal overheads of decoding and matrix loading
2. A waste of on-chip silicon area for any specific personalization, due to the (usually considerable) unused parts of the structure; hence while ROMs may possess supremely high device density, information processing density may be low
3. The possible need in some applications to selectively invert or inhibit one or more of the several output functions (bits)
4. No on-chip latch or clocked register (storage) capability, which may be required in counter and similar applications (see Sec. 3.5)

Thus, for large, fixed combinatorial duties ROMs may be very suitable; however for smaller and more scattered random logic functions, for instance the logic "glue" which is present in many digital systems, the capabilities of the ROM family are not so relevant. Table 3.2 summarizes the size and performance of various available commercial products. Note that in general:

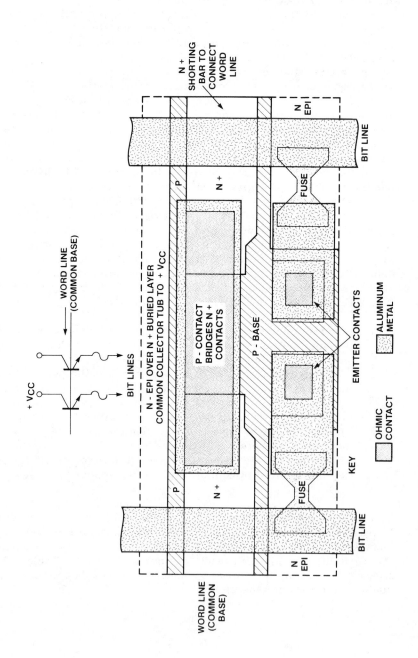

WORD LINE
(COMMON BASE)

+ VCC

BIT LINES

N - EPI OVER N + BURIED LAYER
COMMON COLLECTOR TUB TO + VCC

P - CONTACT
BRIDGES N +
CONTACTS

P - BASE

EMITTER CONTACTS

N +
SHORTING
BAR TO
CONNECT
WORD
LINE

BIT LINE

N
EPI

FUSE

FUSE

BIT LINE

N
EPI

P

N +

P

N +

WORD LINE
(COMMON
BASE)

KEY

OHMIC
CONTACT

ALUMINUM
METAL

132

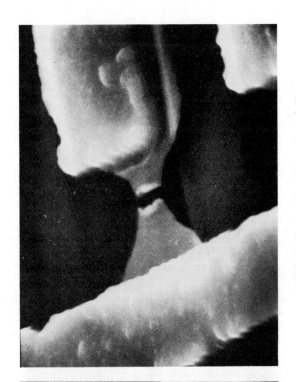

PROGRAMMED FUSE

UNPROGRAMMED FUSE

FIGURE 3.3 Fuse blowing details for a bipolar Schottky TTL programmable device. (Details courtesy of Monolithic Memories, Santa Clara, CA.)

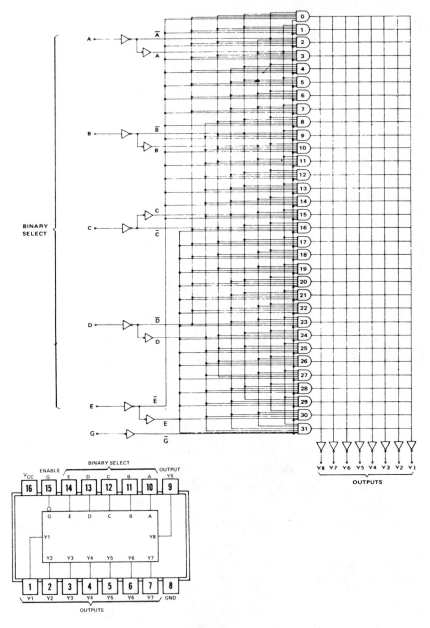

FIGURE 3.4   A SN74188 32 × 8 PROM: (a) logical structure, programming diodes not shown on the minterm array. (b) 16-pin package connections. (Details courtesy Texas Instruments, Bedford, England.)

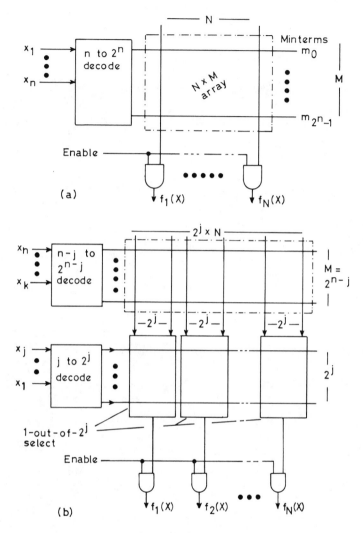

Figure 3.5 Alternative ROM structures: (a) single-decoder architecture as previously shown, suitable for small n. (b) X and Y decoder architecture, necessary for large n to reduce the height M of the programmable array matrix.

TABLE 3.2 Outline Specifications of Various Nonvolatile ROM/PROM Commercial Packages[a]

Type	Organization	Technology	Mask-programmable (ROM)	Field-programmable (PROM)	Read time (ns, approx.)	Package (pins)
74186	64 × 8 ($\frac{1}{4}$K)	TTL	–	Fusible link	50	24
74187	256 × 4 (1K)	TTL	✓	–	40	16
74188	32 × 8 ($\frac{1}{4}$K)	TTL	–	Fusible link	30	16
14524	254 × 4 (1K)	CMOS	✓	–	350	16
2708	1024 × 8 (8K)	MOS	–	UV-erasable	450	24
2532	4096 × 8 (32K)	MOS	–	UV-erasable	450	24
2564	8192 × 8 (64K)	MOS	–	UV-erasable	450	28
2864[b]	8192 × 8 (64K)	MOS	–	Electrically erasable	450	28
5565	8192 × 8 (64K)	MOS	✓	–	150	28
27128	16384 × 8 (128K)	MOS	–	Fusible link	200	28
53128	16384 × 8 (128K)	CMOS	✓	–	250	28
23256	32768 × 8 (256K)	MOS	✓	–	200	28
27256	128K × 8 (1M)	CMOS	✓	–	300	40

[a]Note, × 8 wide devices may be referred to as "1 byte wide."

[b]EEPROM with four multiplexed 2048 × 8 devices in the one package.

1. Mask-programmable MOS ROMs provide the greatest memory capacity per chip, and bipolar field-programmable PROMs the least.
2. Mask-programmable bipolar PROMs provide the fastest access times, with as low as 10 ns being currently reported.
3. Mask-programmable ROMs, whether bipolar or MOS are generally more costly for low volume applications than field-programmable equivalents.

For further details, see available references [1-29].

3.3 PROGRAMMABLE LOGIC ARRAYS

Programmable logic arrays (PLAs) are more directly designed for the realization of random combinatorial logic duties than ROMs. Like ROMs they are invariably multiple input, multiple output structures, but unlike ROMs they do not internally decode the binary input signals down to minterm level. Hence, n-to-2^n decoding is not present.

The basic internal structure of a PLA is shown in Figure 3.6. The binary inputs x_1 to x_n are each internally buffered and also inverted so as to provide x_i and $\bar{x}_i$ from each input, these 2^n signals forming the inputs to the AND array matrix. Product terms may then be programmed by appropriate ANDing, for example the ANDing on a vertical data path through the array of, say, x_1, x_2, and $\bar{x}_4$ would generate the product term $x_1 x_2 \bar{x}_4$. The multiplicity of vertical data paths of the AND array enable multiple product terms to be generated.

The lower programmable array provides the means to selectively OR together the product terms generated by the first array, thus generating each required output $f_1(x)$, $f_2(x)$, . . . in a sum-of-products form. For example, the programming of product terms $x_1 x_2 x_3 + x_2 x_4 + x_1 x_5 x_6$ . . . provides the function

$$f(x) = \{x_1 \bar{x}_2 x_3 + \bar{x}_2 x_4 + \bar{x}_1 x_5 \bar{x}_6 + \ldots\}$$

It should be noted that it is possible to program the AND array to provide minterms, but the total capacity of the AND array is not sufficient to provide all possible 2^n minterms of the n independent input variables. Thus we may have, for example 12 input variables x_1 to x_{12}, an internal capability of 96 programmable product terms, each useful product term containing between 1 (minimum) and 12 (maximum) true or complemented input variables, and eight outputs $f_1(x)$ to $f_8(x)$, packaged in a 24-pin dual-in-line (DIL) package.

While this describes the basic philosophy of all PLA devices, in practice there are many possible internal technological and other

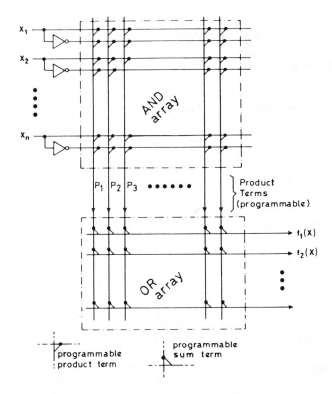

FIGURE 3.6   The fundamental structure of a programmable logic array.
Note that the means of programming the AND and OR matrix connections is similar to ROM and PROM technology.

variations. In theory, all the heirarchy of ROM and PROM devices
given in Table 3.1 could be mirrored in PLA families, and indeed in
the various programmable devices which we will soon cover, but commercial interests have in general restricted the variants. Both mask-
programmable devices (PLAs) and field-programmable devices (FPLAs)
however are available, but erasable/reprogrammable possibilities do
not yet seem to have established any secure commercial market.
    Internal circuit variation from that shown in Figure 3.6, however,
are common. Firstly, as shown in Figure 3.7a, means to selectively
invert each output function can be provided. This may be accomplished
by an EXCLUSIVE-OR gate on each output, one input of which is fed
from logic 0 to noninvert the sum-of-products function, this input
being made logic 1 to provide the complement of the sum-of-products

TABLE 3.3 Outline Specifications of Various PLA/FPLA Commercial Packages

Type	Organization[a]	Technology	Mask- or field-programmable	Access time (ns, approx.)	Package (pin)
74S330	12 × 50 × 6	TTL	Field	30	20
1M5200	14 × 48 × 8	TTL	Field	65	24
82S101	16 × 48 × 8	TTL	Field	50	28
82S201	16 × 48 × 8	TTL	Mask	50	28
93459	16 × 48 × 8	TTL	Field	25	28
82S153	18 × 32 × 10	TTL	Mask	50	20
82S253	18 × 32 × 10	TTL	Field	50	20

[a] 12 × 50 × 6 indicates 12 inputs, 50 product terms, 6 outputs, but not all inputs and output may be simultaneously available if reconfigurable I/Os are present (see Sec. 3.4).

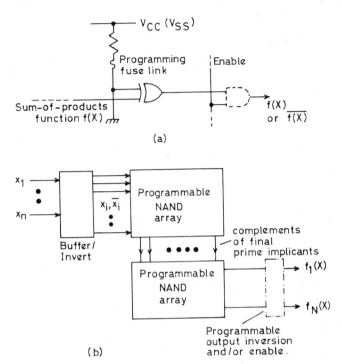

(a)

(b)

FIGURE 3.7   Further PLA/FPLA features: (a) programmable output
inversion using EXCLUSIVE-OR gates as programmable inverters.
(b) NAND/NAND arrays, logically equivalent to AND/OR.

function.[*]   Secondly, rather than the AND/OR arrays previously
shown, it may be more appropriate to make both arrays NAND arrays,
as shown in Figure 3.7b. This has no effect upon the use of the PLA,
since it will be appreciated from basic digital logic theory that, for
example,

$$f(x) = \{x_1\bar{x}_2 + x_3x_4 + \bar{x}_1x_3\}$$

---

*It may be simpler to realize the complement of a required output
function in a sum-of-products form, rather than the true output function. If this is so, then the true output function may be obtained by
a final inversion from the sum-of-products complement. As a trivial
example, the function $f(x) = \bar{x}_1 + \bar{x}_2 + \bar{x}_3$ is most easily realized as

$f(x) = \overline{x_1\bar{x}_2x_3}$.

is equally given by

$$f(x) = \left\{ \overline{(x_1 \overline{x}_2) \cdot (x_3 x_4) \cdot (\overline{x}_1 x_3)} \right\}$$

the latter expression being the equivalent all-NAND form of the former sum-of-products expression. Hence, even if NAND/NAND arrays are present, it is usual to continue to refer to them as AND/OR PLAs.

Table 3.3 outlines details of typical commercial products. The designation 14 × 48 × 8 indicates a PLA with 14 independent binary input variables, 48 internally programmable product lines, and 8 separate output functions. It will be noted that in comparison with ROMs and PROMs, PLAs provide the advantage of a greater possible number of inputs; it will further be appreciated that some of the output functions may be functions of a subset of the input variables, while other functions may be functions of separate sets of the input variables. This may be achieved with ROM structures, but is particularly appropriate with PLAs. A further advantage with PLAs is that static race hazards in outputs may be eliminated by appropriate prime implicant coverage [2, 6, 30], such a facility being entirely absent in the minterm ROM and PROM structures.

The disadvantages and restriction with PLAs are:

1. It is necessary for the logic designer to minimize the required output functions to sum-of-products form, in which the total number of prime implicants is within the capacity of the PLA product array; absolute individual minimization of each output function may not be necessary, provided the total lies within the available array capacity.*

2. Should the required output functions be functions which do not classically minimize into compact prime implicant form, but instead have a "scattered" minterm structure (e.g. multiple-input EXCLUSIVE-OR functions), then the minterm capability of the ROM (PROM) may be more appropriate than a PLA.

We may conclude that somewhat more initial logic design effort may be required with PLAs compared with ROM truth table realizations, but this is well within the capabilities of any professional logic designer.

---

*Indeed individual output function minimization does not always provide the minimum total number of prime implicants, due to the possibility of sharing product terms between outputs. However, the optimum realization of multiple output functions, short of exhaustive search procedures, remains a generally unsolvable logic problem.

For further design and application details, see the many available references [1—4, 6, 8, 23, 30—40, 46].

3.4 PROGRAMMABLE ARRAY LOGIC; PROGRAMMABLE GATE ARRAYS

Certain simpler forms of PLAs are also found on the commercial market, which may have advantages for particular random logic applications. Currently these alternatives are most widely available in field-programmable form, although the abbreviations used for them may not always make this clear.

Programmable array logic (PAL) devices are field-programmable devices with a *programmable* AND array followed by a *fixed* OR assembly.* As in PLAs this AND/OR combination may in reality be NAND/NAND, but we will here continue to refer to such assemblies as AND/OR without loss of generality. Figure 3.8 gives the general architecture of such devices.

A further complication with commercial designations in this area of programmable AND/fixed OR devices is that some products marketed under the general designation programmable array logic contain clocked bistable circuits within the same package. We will here refer to such augmented products by the alternative designation PLS, programmable logic sequencers, (PLS), as covered in the following section. Finally, certain PAL products are obtainable in mask-programmable form, directly interchangeable with their field-programmable counterparts, being termed hard-array logic (HAL) devices [41, 42]. However, we will continue in this section to consider the basic architecture as given in Figure 3.8.

The capacity of a PAL may be given in the same manner as that used for a PLA, namely [(number of dedicated binary input) × (number of available product terms) × (number of output functions per package)]. However, most commercial products have a larger programmable AND array than this suggests, since a number of the output circuits may be reconfigured to act either as outputs or as additional inputs, as required. Figure 3.9 illustrates a particular commercial product. It may be noted that such dual-purpose input/output terminals may also be found on certain FPLAs [32, 39], although this is not a universal feature.

Programmable gate arrays (PGAs) are similar to the above PALS, but do not contain any OR capability to generate directly sum-of-product

---

*PAL is a registered trademark of Monolithic Memories, Inc., but has obtained wider general usage.

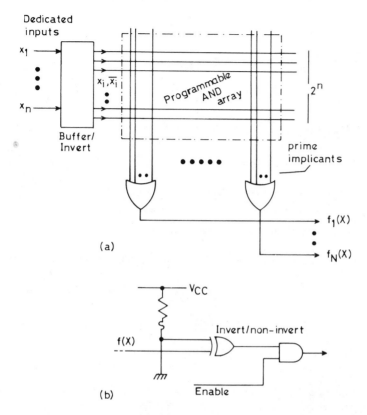

FIGURE 3.8 A basic PAL structure: (a) programmable AND array, fixed output OR assembly. (b) Further output facilities.

output functions. They are, therefore, basically prime implicant (AND) generators.

Figure 3.10 shows the basic structure of such devices. The AND matrix is field-programmable in the usual manner, with programmable output inversion and Enable of each product term possibly being provided.

However, like the PAL, the PGA is usually provided with some or all of its outputs reconfigurable as inputs. Figure 3.11 illustrates a particular commercial product, in which any of its 12 output pins may be reconfigured as inputs additional to the six fixed (dedicated) inputs. Further control facilities will also be noted on this particular family of devices [32]. The programming of output function $I_0 I_1 \bar{I}_5$ on O/P B_{11}, and function $\bar{I}_0 I_1 I_5$ on O/P B_{10} is indicated in this diagram.

A final feature concerning the logic capability which the additional inputs to the AND array provide may also be observed. Taking for

Functional Diagrams

HPL-77319/16LE8

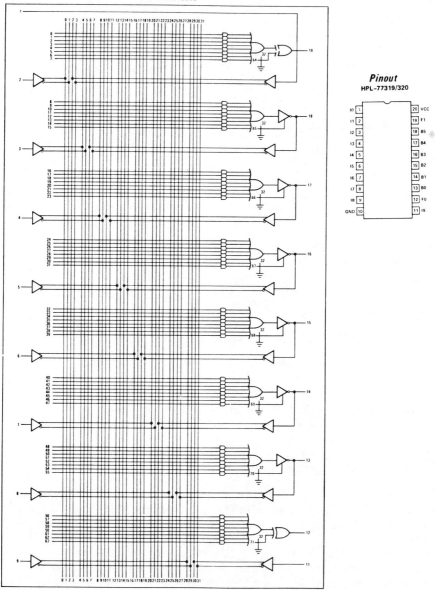

Pinout
HPL-77319/320

FIGURE 3.9 A commercial field-programmable PAL package, with 10 dedicated inputs, 32 programmable product terms, 8 sum-of-product terms, 2 outputs programmable true or inverted, 6 additional outputs reconfigurable as inputs if required. V_{CC} = 5 V, propagation delay = 35 ns. (Details courtesy Harris Corp., Melbourne, FL.)

144

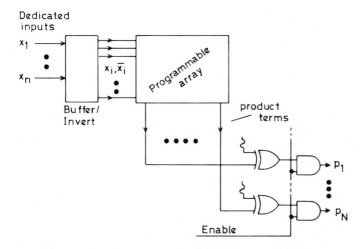

FIGURE 3.10   A basic PGA structure. Note the programmable AND but no OR facility.

example the FPGA structure of Figure 3.11, if product term $\overline{x}_1 x_2 \overline{x}_3$ is generated by the AND array from the dedicated inputs, then this term or its complement may be connected back into the AND array and combined with further variables or further product terms. If we AND the complement of $\overline{x}_1 x_2 \overline{x}_3$ with $x_4 x_5$, we then obtain the function

$$f(x) \;=\; x_4 x_5\,\overline{(\overline{x}_1 x_2 \overline{x}_3)}$$

$$=\; x_4 x_5 (x_1 + \overline{x}_2 + x_3)$$

$$=\; x_1 x_4 x_4 + \overline{x}_2 x_4 x_5 + x_3 x_4 x_5$$

Thus a sum-of-products expression may be realized without directly employing an OR in the architecture of the programmable device. A disadvantage of the technique, however, is that propagation times may be longer than desirable, and glitches may be accentuated in final output functions. For further PAL and PGA details and applications, see various available publications [1, 22, 23, 31, 32, 39–43].

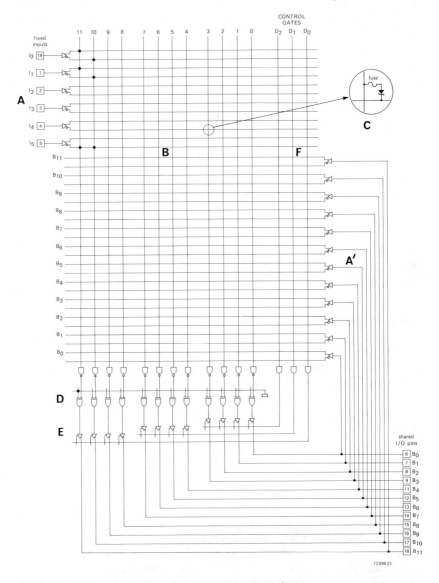

FIGURE 3.11   A commercial Schottky TTL field-programmable gate-array (FPGA).   A = dedicated inputs (6).   A' = additional inputs (12). B = programmable product array (18 × 12).   C = fuse programmable detail.   D = programmable inversions (12).   E = programmable I/Os (12).   F = programmable control matrix. V_{CC} = 5 V, propagation delay = 30 ns.   (Details courtesy Philips, Eindhoven, the Netherlands.)

3.5 PROGRAMMABLE LOGIC SEQUENCERS

None of the programmable devices so far illustrated have contained any bistable circuits or other register elements; all have been fundamentally combinatorial devices, their present-state outputs being controlled by the state of the input data, assuming no closed feedback loops have been incorporated in their personalization.

However, since NAND functions are available from all programmable devices, then it is theoretically possible to build up any desired synchronous or asynchronous network, due to the universality of NAND primitives. Simple latch (Set – Reset) circuits may readily be realized by two cross-coupled NAND gates, but the construction of more sophisticated logic elements, such as clocked type JK circuits, would be tedious.

Therefore, the availability of more powerful clocked bistable circuit elements within programmable LSI packages has obvious advantages; it enables such devices to be readily employed as synchronous code generators, counters, and the like, limited only by the amount of storage and combinatorial logic available per package.

The programmable devices which contain such a facility are best termed programmable logic sequencers (PLS). The currently available ones are generally field-programmable, and therefore the designation FPLS is widely applied. An exception to this is the commercial family of PAL devices, (see the preceding section, p. 142), which contain storage and may also be mask-programmable; the previously noted designation HAL is also applied to these latter variants [41, 42].

The general structure of a PLS (FPLS) is shown in Figure 3.12. The programmable AND array is similar to PLAs, PALs, and PGAs, the product output terms from which are available either as combinatorial output functions as before, or as internal clock-steering (data) inputs for the clocked bistable elements.

There are, however, several possible variants on this general theme. Firstly, the product terms from the AND array may feed into a fixed OR assembly to generate the required internal sum-of-products logic terms [41, 42], or may feed a programmable OR array as in a normal PLA structure [32, 43]. Additionally, the clocked bistable elements may be type D or type JK, the latter being more universal but of greater complexity. As far as is known no commercial PLS has direct-set (preset) or direct-reset (clear) provision on the registers, the only means of control being via the clock and clock-steering logic. Figure 3.12b and c illustrate these variants, together with feedback into the AND matrix which is invariably provided for clock-steering purposes. Reconfigurable combinatorial outputs (see Fig. 3.11), may also be provided.

Figure 3.13 illustrates a commercial FPLS product, fabricated in bipolar technology and packaged in a 24-pin dual-in-line package. An

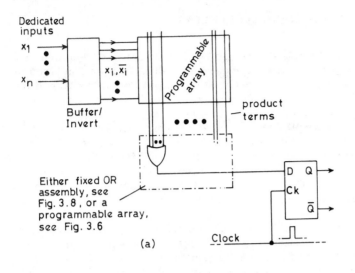

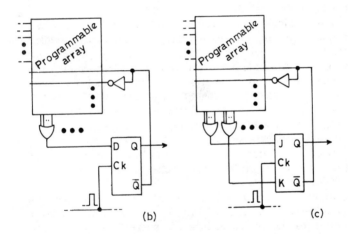

FIGURE 3.12   A basic PLS structure: (a) outline structure; (b) storage feedback into the programmable AND array; (c) type JK storage instead of type D.

interchangeable mask-programmable version of this particular product
is also available, thus enabling prototype system design to be undertaken by an OEM on the field-programmable version, the proven requirements being finally personalized in mask-programmable form. Additional logic capability will be noticed in the D input clock-steering
signals in the design over and above the basic concepts shown in Figure 3.11. For further details of the internal architecture and the use
of these devices, see various available publications [1,6,22,23,31,32,
41−43].

3.6 OTHER PROGRAMMABLE STRUCTURES

The programmable structures which we have considered in the previous
sections of this chapter, whether vendor-mask-programmable or user-
field-programmable, have all being variants on the classic sum-of-
products logic structure. All involve the generation of product (AND)
terms from the binary input variables, and a final summation (OR) of
these terms to generate appropriate outputs. In the case of the ROM
(PROM) the product terms are fixed, being the 2^n minterms of the n
input variables, but in the remaining cases, programmable product
terms with up to n true or complemented input variables have been
shown to be available.

Table 3.4 summarizes these families, and the programmability of
each. All are currently available in field-programmable form, but not
all, as far as is known, are available in mask-programmable form. Similarly, all families are available in bipolar technology, but in general
only the ROM/PROM and PLA/FPLA families are also available in MOS
or CMOS. This commercial status is subject to continuous revision and
evolution.

However, while the families of devices listed in Table 3.4 constitute
the most widely known range of programmable devices, mention should
be made of certain others which are commercially available. These
include programmable diode matrices and programmable multiplexers.

Programmable diode matrices (PDMs or DMs) constitute the simplest
possible useful programmable device. They consist merely of a matrix
array of orthogonal connections, each crossing point linked by a diode
plus a fusible link as employed in many other programmable arrays.
No additional logic facilities are present.

Figure 3.12 indicates the general structure of such simple devices,
which are available in several matrix sizes. The field programming of
DMs is particularly simple, since all that is required is the selection
of an appropriate column terminal and an appropriate row terminal on
the package connections, fusing current then being passed down the
selected column, through the fuse link and diode, and along the selected
row to open-circuit this path.

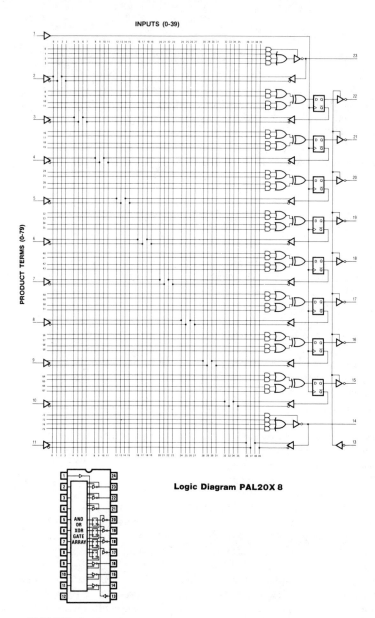

INPUTS (0-39)

PRODUCT TERMS (0-79)

Logic Diagram PAL20X 8

FIGURE 3.13 A commercial Schottky TTL field-programmable logic sequencer (FPLS): dedicated inputs 10, additional array inputs 10, programmable AND array 20 × 76, fixed OR/EXCLUSIVE-ORs 10, type D storage and outputs 8, reconfigurable combinatorial I/Os 2, V_{CC} = 5 V, propagation delay clock to output or feedback 30 ns. (Details courtesy Monolithic Memories, Santa Clara, CA; mask-programmable equivalent = HAL20X8.)

TABLE 3.4 Summary of the Principal Families of Programmable Devices

Family	Product (AND) terms	Summation (OR)	Reconfigurable I/Os	Bipolar or MOS/CMOS	Mask- or field-programmable	Additional facilities
ROM, PROM	Fixed (minterms)	Programmable	No	Both	Both	–
PLA, FPLA	Programmable	Programmable	Not usually	Both	Both	–
PAL	Programmable	Fixed	Yes	Bipolar[a]	Field	Also available with storage, see text, in which case they may more properly be termed PAL-FPLS or HAL-PLS
HAL	Programmable	Fixed	Yes	Bipolar[a]	Mask	
PLS, FPLS	Programmable	Programmable or fixed	Yes	Bipolar[a]	Both	Type D or type JK clocked storage facilities built in, with programmable clock-steering
PGS	Programmable	Not present	Yes	Bipolar[a]	Field[a]	–

[a]Status as far as is generally available; further variants may be or may become available.

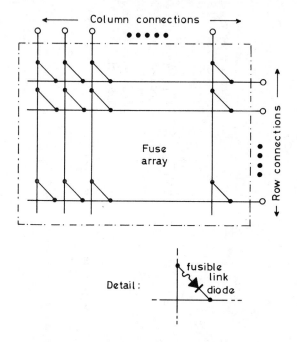

FIGURE 3.14   The structure of a simple programmable diode matrix (PDM).

Available DMs are generally small devices, for example 8 × 6 or 10 × 4 arrays packaged in 14-pin dual-in-line packages. They find potential use in simple encoding or decoding applications, addressing networks, or where simple diode isolation is required.

Finally, programmable multiplexer (PMUX) devices. These basically consist of a n × M programmable fuse array, which serves to connect n-dedicated inputs selectively to M array output lines, followed by a M-to-1 line multiplexer. This is shown in Figure 3.15a. In general all but one diode will be programmed open-circuit in each of the array output lines, so as connect only the desired input to the required array output line. The multiplexer then multiplexes these configured input signals to the final output under the control of the m data-select inputs. Note that $M = 2^m$, as is usual in all multiplexers.

A practical PMUX will contain more than one such assembly, as illustrated in Figure 3.15b. Final buffering and/or Enable of each multiplexed output is provided. Further details may be found in published literature [8,23,31].

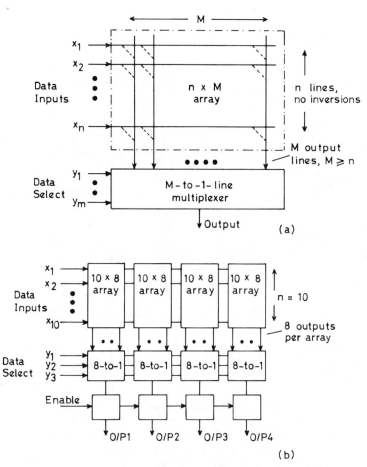

FIGURE 3.15   The structure of a programmable multiplexer (PMUX).
(a) Basic structure for each output. (b) Typical PMUX architecture
for a 20-pin DIL package.

3.7 APPLICATIONS AND PERSONALIZATION

It will be apparent from our discussions that programmable LSI devices
do not have the internal logic capacity to realize very complex logic
systems using one package. In general the number of dedicated inputs
is around the 10 mark, and the number of available outputs about 8.
In the case of programmable logic sequencers, the number of internal
clock storage elements is possibly 4–8. Thus, systems involving tens
of variables, hundreds of product terms, and many state variables

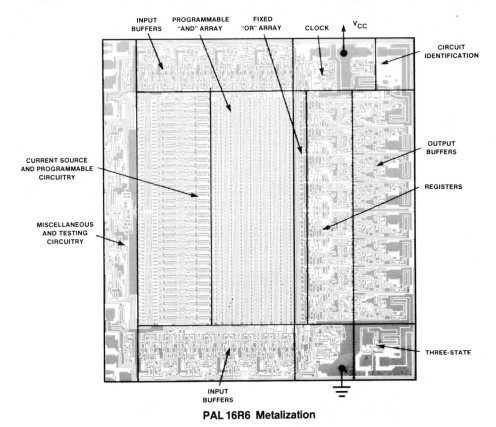

INPUT   PROGRAMMABLE   FIXED
BUFFERS   "AND" ARRAY   "OR" ARRAY   CLOCK V_{CC}

CIRCUIT
IDENTIFICATION

CURRENT SOURCE
AND PROGRAMMABLE
CIRCUITRY

OUTPUT
BUFFERS

REGISTERS

MISCELLANEOUS
AND TESTING
CIRCUITRY

THREE-STATE

INPUT
BUFFERS

PAL 16R6 Metalization

FIGURE 3.16   Microphotograph of a commercial programmable logic
sequencer, switching characteristics comparable with 7400-series TTL.
(Courtesy Monolithic Memories, Santa Clara, CA.)

cannot be accommodated within one such LSI package, as may be possible if other custom or semicustom routes are followed.

Nevertheless, there are many industrial and other applications which
do not involve large systems for which the programmable device may
be particularly suitable. For example, vending machines, small security
alarm systems with four to six sensor inputs, control sequencers for
small automated plants or equipment, electronic game items and the like
are potential candidates for their use. Figure 3.16 illustrates a commercial product, which indicates the professionalism available within
such commercial devices.

TABLE 3.5   Considerations in the Adoption of Programmable
Devices

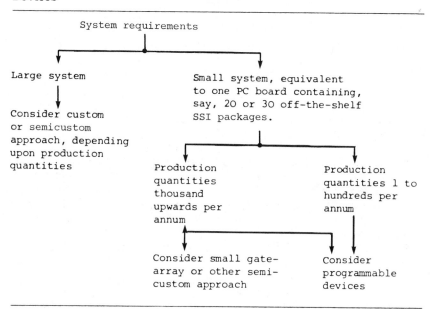

The market size for the product is also significant. Should production quantities of tens of thousands be involved, the alternative
semicustom approaches of the following chapters would be more realistic, but for small quantity requirements one may profitably consider
programmable alternatives. These various considerations may be broadly summarized as in Table 3.5. The added question of the long-term
reliability of programmable LSI devices, particularly field-programmable
versus nonprogrammable devices, should also receive consideration,
this being an ongoing and evolving situation [26–28,41].

The performance specification of programmable devices will usually
be appropriate, and not at all restrictive. We have noted previously a
typical 30 ns propagation delay for bipolar programmable devices, which
is somewhat slower than that available from off-the-shelf 7400 series
TTL packages. It may be appreciated that with off-the-shelf TTL
packages it is often necessary to have several levels of gating in a
combinatorial output function, due to fan-in limitations on available
gates; with programmable devices, fan-in capability is generally greater, and hence two-level realizations of large sum-of-products functions
are more readily available. For example, see the following [32]:

No. of inputs in required function	TTL gates in cascade using SSI packages	Propagation delay (ns)		
		TTL 74 series	TTL 74LS series	Bipolar programmable
<8	1	20	17	30
>8	2	40	34	30

Additionally, the interpackage wiring capacity of small-scale integration (SSI) packages assembled on a printed circuit (PC) board always degrades the maximum available performance, and any transfer
of interconnections to on-chip has advantages. Further performance
details and tradeoffs may be found in Muroga [1] and elsewhere [32–
42]. Programmable devices will not, however, quite match the performance capability of comparable semicustom, particularly cell-library
realizations.

System design using programmable devices is fundamentally no
different than any other means of hardware realization. A certain
ingenuity in partitioning a given requirement within the particular
limitations of a given programmable device may be involved, but this
is within the normal activities of an original equipment manufacturer
(OEM) system designer. The minimization of several required combinatorial functions from a common set of input variables has already
been remarked upon (Sec. 3.3), but again, in general no problems
beyond the capabilities and intuitive abilities of the designer will be
encountered.

Having formulated the system requirements, the following steps
then follow:

1. Data preparation, where the particular design requirements
    are transferred to a connection listing for the programmable
    device to be used
2. Personalization of the device, either mask or field
3. Functional test, to confirm the correct operation of the per-
    sonalized device

The first stage in this procedure normally involves some form of
"programming map," which lists the required on-chip connections.
Such a map, sometimes termed a "memory map" or a "personality table,"
will have a matrix layout which mirrors the physical on-chip layout,
thus defining which links have to be mask-programmed in or field-
programmed out.

Figure 3.17 gives a map for a 32 × 8 PROM illustrated in Figure
3.4, with the programming of the 256 points required to implement a

CIRCUIT TYPES SN5488A, SN7488A
256-BIT READ-ONLY MEMORIES

TRUTH TABLE/ORDER BLANK

CUSTOMER _____

PURCHASE ORDER NO. _____

CUSTOMER PART NO. _____

THIS PORTION TO BE COMPLETED BY TI

TI PART NO.: _____

S.O. NO.: _____

DATE RECEIVED: _____

WORD	INPUTS						OUTPUTS							
	BINARY SELECT					ENABLE								
	E	D	C	B	A	G	Y8	Y7	Y6	Y5	Y4	Y3	Y2	Y1
0	L	L	L	L	L	L	H	H	H	H	H	H	H	H
1	L	L	L	L	H	L	H	H	H	L	H	H	L	L
2	L	L	L	H	L	L	H	H	H	L	H	L	H	L
3	L	L	L	H	H	L	H	H	H	H	H	L	L	H
4	L	L	H	L	L	L	H	H	H	L	H	L	L	H
5	L	L	H	L	H	L	H	H	H	H	H	L	H	L
6	L	L	H	H	L	L	H	H	H	H	H	H	L	L
7	L	L	H	H	H	L	H	H	H	L	H	H	H	H
8	L	H	L	L	L	L	H	H	H	H	H	L	L	L
9	L	H	L	L	H	L	H	H	H	L	H	L	H	H
10	L	H	L	H	L	L	H	H	H	L	H	H	L	H
11	L	H	L	H	H	L	H	H	H	H	H	H	H	L
12	L	H	H	L	L	L	H	H	H	L	H	H	H	L
13	L	H	H	L	H	L	H	H	H	H	H	H	L	H
14	L	H	H	H	L	L	H	H	H	H	H	L	H	H
15	L	H	H	H	H	L	H	H	H	L	H	L	L	L
16	H	L	L	L	L	L	H	H	H	L	L	H	H	L
17	H	L	L	L	H	L	H	H	H	H	L	H	L	H
18	H	L	L	H	L	L	H	H	H	H	L	L	H	H
19	H	L	L	H	H	L	H	H	H	L	L	L	L	L
20	H	L	H	L	L	L	H	H	H	H	L	L	L	L
21	H	L	H	L	H	L	H	H	H	L	L	L	H	H
22	H	L	H	H	L	L	H	H	H	L	L	H	L	H
23	H	L	H	H	H	L	H	H	H	H	L	H	H	L
24	H	H	L	L	L	L	H	H	H	L	L	L	L	H
25	H	H	L	L	H	L	H	H	H	H	L	L	H	L
26	H	H	L	H	L	L	H	H	H	H	L	H	L	L
27	H	H	L	H	H	L	H	H	H	L	L	H	H	H
28	H	H	H	L	L	L	H	H	H	H	L	H	H	H
29	H	H	H	L	H	L	H	H	H	L	L	H	L	L
30	H	H	H	H	L	L	H	H	H	L	L	L	H	L
31	H	H	H	H	H	L	H	H	H	H	L	L	L	H
ALL	X	X	X	X	X	H	H	H	H	H	H	H	H	H

COMPLETE THE LOGIC DESIRED FOR 256 BITS.
INDICATE H FOR HIGH LEVEL OR L FOR LOW LEVEL.

H = high level, L = low level, X = irrelevant

FIGURE 3.17 Programming details for a SN7488 ROM to implement a Hamming code transmitter. (Note: LH array is the fixed minterm generation from inputs A to E, RH array is the required OR personalization of the eight outputs Y_1 to Y_8.) (Details courtesy Texas Instruments, Bedford, England.)

specific product. Such a table forms the input specification for the
mask- or field-personalization procedure (see below).

Possibly the most complex design and programming specification is
present in programmable logic sequencers, particularly if it is required
to be able to test the final combinatorial and sequential logic in the
most expiditious manner.* Figure 3.18a shows details of a fuse-
programmable Schottky TTL package, with four type JK circuits which
may be reconfigured as type T or type D when required. Figure 3.18b
gives the programming details for this package when realizing a reversible counter/shift register plus a binary multiplier. Further details relating to the steps in this particular design may be found published [32].

The personalization of mask-programmable devices in accordance
with the programming details is the province of the vendor. The OEMs
responsibility ceases with the compilation of the programming map or
equivalent data, although the vendor may return some software print-
out before mask programming any devices, to cross-check the customer's design data.

For field-programmable devices, however, the OEM may be actively
involved in considerably greater sophistication, should he undertake
his own device personalization. For simple PROM programming, it may
be adequate to employ a simple stand-alone hardware unit into which
the programming requirements may be manually entered via a keyboard, such as illustrated in Figure 3.19. Such a unit will generally
consist of the keyboard to enter the programming details, RAM to store
this data entry, some means to check and edit the data before executing
the device programming, an appropriate interface unit between the
data and the device to be programmed, a socket adaptor to take the
device being programmed, and control electronics to finally execute
the selective on-chip fuse-blowing schedule.

Appreciable error prevention and circuit verification is normally
present, usually including verification that the device to be programmed
is initially blank, continuity checks on the device being programmed,
a sum check of the number of links to be fused and those already fused,
and other facilities. The hardware interface circuit which controls the
electrical fusing signals, sometimes termed the "personality unit," has
specific output signals for each type of device to be programmed, often
being an interchangeable plug-in module to suit a particular range of
fuse-programmable devices.

---

*Fast testing techniques may require the reconfiguration of sequen-
tial networks into shift register or other configurations in order to be
able to exercise all the logic without the need to apply an excessive
number of clock pulses.

Further facilities are frequently available, for example, a simple interface to an external source of data, such as a paper-tape recorder or computer output bus. Alternatively, or in addition, the unit may be loaded with data from a preprogrammed master PROM.

More sophisticated programming units may be employed, particularly where FPLA and FPLS devices are involved [8,44]. Such units may become small stand-alone computer-aided design (CAD) workstations, since software programs to complete and verify the system design may be incorporated. A particular feature may, for example, be the min-imization of the product terms for the multiple output functions of a PLA, and possibly test vector generation [8,39,41,44,45].

Table 3.6 gives the flow chart of a possible CAD software program. Such software is usually generated by the manufacturers of the pro-gramming units, and made available to OEMs to use either in house, or remotely via a telephone link.

Figure 3.20 illustrates a unit with FPLA and FPLS capabilities. A necessary feature of all the more sophisticated units is the provision of a video display unit (VDU) to provide the visual interface between the operator and the controlling software of the unit. Plug-in per-sonality modules specific to particular programmable devices are also shown. Later units may not have such plug-in modules, but instead may have a fixed hardware interface whose output signals are con-trolled by the unit software.

A further aspect concerning the programming and testing of field-programmable devices must be noted. With simple PROM devices, con-sisting of only one fuse-programmable array, the verification that each individual fuse in the array has been correctly blown is simultaneous with a check of the required input minterm/output logic, the fixed in-put decoding being part of the fuse selection procedure. However, with FPLAs and similar devices which contain more than one fuse-programmable array, the individual fuse blowing and testing is selec-tive to each array; verification of each fuse does not now consitiute an overall functional test of the personalized device, since some further on-chip defects may be present in the remaining logic and interconnect [39].

Thus, functional testing as well as fuse-blowing verification is nec-essary. The latter is straightforward, being an integral part of the selective fuse-blowing procedure, but the former is the normal and sometimes difficult problem of functional testing of LSI devices. The more complex available programmable devices become, the more diffi-cult full functional testing may be, particularly where internal sequen-tial networks are involved.

Finally, mention may be made of an additional facility which may be present in order to increase the commercial security of a personalized device in an original equipment manufacturer's product. Normally the internal circuitry used for fuse-blowing and subsequent verification

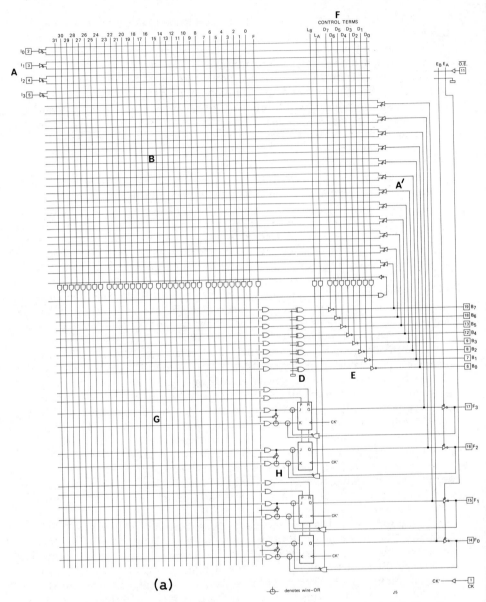

(a)

denotes wire-OR

FIGURE 3.18   Detail of the architecture and programming details of a large programmable logic sequencer. (a) Circuit details, (b) programming table details for a reversible counter plus multiplier. (Details courtesy Philips, Eindhoven, the Netherlands.)

can be used at any time to determine the logic patterns stored within the programmable array(s). However, it is possible to obtain devices which contain a "last fuse," final blowing of which disables the internal array verification circuitry, and prevents ready copying of the proprietary design [41]. Other detailed features may be present, or will appear as further programmable devices evolve.

(b)

FIGURE 3.18 (Continued)

The programmable logic array program table (Figure 3.18b). The table contains the following principal regions:

F/F TYPE — J/K, D

$E_B = A$, **$E_A = A$** with entries A A A A

POLARITY — H

Left-hand descriptor columns:
- $E_A \cdot E_B$: IDLE, CONTROL, ENABLE, DISABLE
- F/F TYPE: J/K, D (OR)
- (Q = J/K): TOGGLE, SET, RESET, HOLD (OR); HIGH, LOW (POL.)
- C: INACTIVE, GENERATE, PROPAGATE, TRANSPARENT (AND); P, R, B(O), (Q = D) ACTIVE, INACTIVE (OR)
- PROGRAM TABLE ENTRIES: I, B(I), Q(P) INACTIVE, I·B·Q, I·B̄·Q̄ DON'T CARE (AND)

TERM	C	I 3	2	1	0	B(I) 7	6	5	4	3	2	1	0	Q(P) 3	2	1	0	Q(N) 3	2	1	0	P B	A	R B	A	B(O) 7	6	5	4	3	2	1	0	POL
0	•	L	H	L	H	L	–	–	–	–	–	–	–	–	–	–	–	–	–	–	0	–	–	–	–	–	–	–	–	–	–	–	–	A
1	•	L	H	L	H	L	–	–	–	–	–	–	–	–	–	–	L	–	–	0	–	–	–	–	–	–	–	–	–	–	–	–	–	A
2	•	L	H	L	H	L	–	–	–	–	–	–	–	–	–	L	L	–	0	–	–	–	–	–	–	–	–	–	–	–	–	–	–	A
3	•	L	H	L	H	L	–	–	–	–	–	–	–	L	L	L		0	–	–	–	–	–	–	–	–	–	–	–	–	–	–	–	A
4	•	L	L	H	H	L	–	–	–	–	–	–	–	–	–	–	–	–	–	–	0	–	–	–	–	–	–	–	–	–	–	–	–	A
5	•	L	L	H	H	L	–	–	–	–	–	–	–	–	–	–	H	–	–	0	–	–	–	–	–	–	–	–	–	–	–	–	–	A
6	•	L	L	H	H	L	–	–	–	–	–	–	–	–	H	H		–	0	–	–	–	–	–	–	–	–	–	–	–	–	–	A
7	•	L	L	H	H	L	–	–	–	–	–	–	–	H	H	H		0	–	–	–	–	–	–	–	–	–	–	–	–	–	–	A
8	•	L	H	L	L	L	–	–	H	L	–	–	–	H	–	–		–	–	–	H	–	–	–	–	–	–	–	–	–	–	–	A
9	•	L	H	L	L	L	–	–	H	H	–	–	–	H	–	–		–	–	–	H	–	–	–	–	–	–	–	–	–	–	–	A
10	•	L	H	L	L	L	–	–	L	H	–	–	–	–	H	–		–	–	–	H	–	–	–	–	–	–	–	–	–	–	–	A
11	•	L	H	L	L	L	–	–	L	H	–	–	–	–	–	H		–	–	H	–	–	–	–	–	–	–	–	–	–	–	–	A
12	•	L	H	L	L	L	–	–	H	L	–	–	H	–			–	–	H	–	–	–	–	–	–	–	–	–	–	–	–	–	A
13	•	L	H	L	L	L	–	–	H	H	–	–	H	–			–	–	H	–	–	–	–	–	–	–	–	–	–	–	–	–	A
14	•	L	H	L	L	L	–	–	L	H	–	–	–	H	–		–	H	–	–	–	–	–	–	–	–	–	–	–	–	–	A
15	•	L	H	L	L	L	–	–	H	L	–	–	–	H		–	H	–	–	–	–	–	–	–	–	–	–	–	–	–	A
16	•	L	H	L	L	L	–	–	H	H	–	H	–	–		–	H	–	–	–	–	–	–	–	–	–	–	–	–	–	A
17	•	L	H	L	L	L	–	–	L	H	–	–	H	–	H	–	–	–	–	–	–	–	–	–	–	–	–	–	–	A
18	•	L	H	L	L	L	–	–	H	L	–	–	H	–	H	–	–	–	–	–	–	–	–	–	–	–	–	–	–	A
19	•	L	H	L	L	L	–	–	H	H	–	–	H	H	–	–	–	–	–	–	–	–	–	–	–	–	–	–	A
20	•	L	L	H	L	L	–	–	–	–	–	–	–	–	H	–	–	–	H	–	–	–	–	–	–	–	–	–	–	A
21	•	L	L	H	L	L	–	–	–	–	–	–	H	–	–	–	H	–	–	–	–	–	–	–	–	–	–	–	A
22	•	L	L	H	L	L	–	–	–	–	–	H	–	–	–	H	–	–	–	–	–	–	–	–	–	–	–	–	A
23	•	L	L	H	L	L	–	–	–	–	–	–	H	H	–	–	–	–	–	–	–	–	–	–	–	–	–	A
24	•	H	–	–	–	L	–	–	–	–	–	–	–	–	–	–	–	–	–	–	–	–	–	–	–	–	–	–	•
25	•	–	–	–	–	H	–	–	–	–	–	–	–	–	–	–	–	–	–	–	–	–	–	–	–	–	–	–	•
26	A	–	–	–	–	–	L	L	–	–	–	–	–	–	–	–	–	–	–	–	–	–	–	–	–	–	–	•
27																																		
28																																		
29																																		
30																																		
31																																		
F		–	–	–	L	–	–	–	–	–	–	–	–	–	–	–	–																

	15–0 (D registers)
LB	
LA	
D7	0 0 0 0 0 0 0 0 0 0 0 0 0 0 0 0
D6	0 0 0 0 0 0 0 0 0 0 0 0 0 0 0 0
D5	0 0 0 0 0 0 0 0 0 0 0 0 0 0 0 0
D4	0 0 0 0 0 0 0 0 0 0 0 0 0 0 0 0
D3	0 0 0 0 0 0 0 0 0 0 0 0 0 0 0 0
D2	
D1	
D0	– – – – – – L – – – – – – –

PIN NO.	5	4	3	2	19	18	13	12	9	8	7	6	17	16	15	14
VARIABLE NAME	INIT	LEFT/DOWN	RIGHT/UP	COUNT/SHIFT	RESET	H̄ BUSY	Y	X					ACTIVE	D	C	B	A

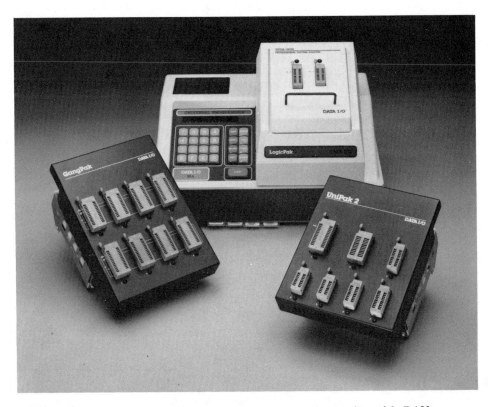

FIGURE 3.19 A commercial portable programming unit, with RAM
buffer memory, hexadecimal keyboard entry, direct readout of RAM
or device at any location, error-prevention routines, plug-in programming adaptor. (Photo courtesy DATA I/O, Redmond, WA.)

TABLE 3.6   Flow Chart of a Possible CAD Software
Routine for the Design and Personalization of FPLAs,
etc.

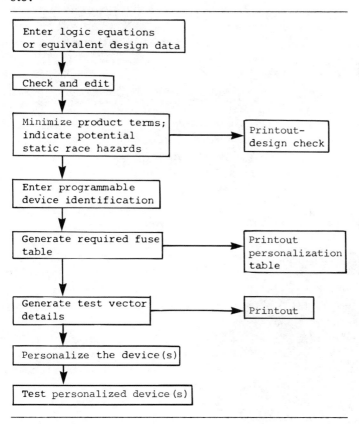

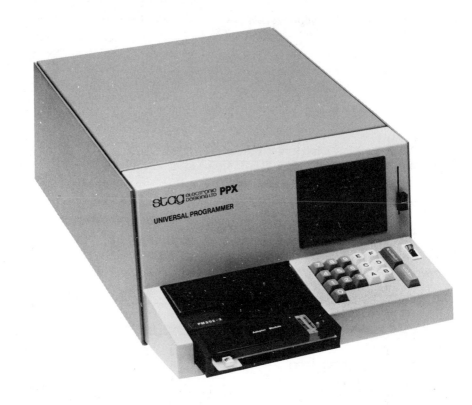

FIGURE 3.20 A commercial universal programming unit, with FPLA and FPGA, etc. programming capabilities; facilities include Boolean entry of data, VDU menu, full-check facilities. (Photo courtesy Stag Electronic Designs Ltd., Welwyn Garden City, England.)

3.8 SUMMARY

Programmable devices certainly play a significant part in the range of options available to an original equipment designer for the hardware realization of his particular product. As indicated in Table 3.5, the size of the system for which programmable devices may be appropriate is limited, but nevertheless a considerable range of simple domestic and industrial products lie within such limits. For this reason, we have felt it appropriate to discuss these products in some detail within this text, along with the more generally accepted forms of realization for custom and semicustom microelectronics. As in the whole area of custom and semicustom microelectronics, the fundamental parameters which the OEM designer must consider are the required system complexity and the quantity to be manufactured.

REFERENCES

1. Muroga, S. *VLSI System Design*. Wiley-Interscience, New York, 1982.
2. Cahill, S. J. *Digital and Microprocessor Engineering*. Ellis Harwood, UK and John Wiley, New York, 1982.
3. Mavor, J., Jack, M. A., and Denyer, P. B. *Introduction to MOS LSI Design*. Addison-Wesley, London, 1983.
4. Blakeslee, T. R. *Digital Design with Standard MSI and LSI*. Wiley-Interscience, New York, 1975.
5. Agajanian, A. H. *MOSFET Technologies: A comprehensive Bibliography*. Plenum Press, New York, 1980.
6. Fletcher, W. I. *An Engineering Approach to Digital Design*. Prentice-Hall, Englewood Cliffs, NJ, 1980.
7. Intel Inc. *The Semiconductor Memory Book*. John Wiley, New York, 1978.
8. Data I/O. How to survive in the programming jungle. Data I/O Corp, Issaquah, WA, May 1978.
9. Wyland, D. C. Using PROM's as logic elements. *Computer Design.*, 13:98-100, 1974.
10. McDowell, J. Large bipolar ROM's and PROM's revolutionise logic design. *Computer Design*, 13:100-104, 1974.
11. Ludwig, J. A. A 50 Kbit Schottky cell bipolar read-only memory. *IEEE Journal Solid-State Circuits*, SC.15:816-820, 1980.
12. Wilson D. R., and Schroeder, P. R. A 100 ns 150 mW 64 K bit ROM. *Proc. IEEE Int. Solid-State Circuits Conf.*, pp. 152, 153, 1978.
13. ROM's and PROM's are moving to greater densities. *Electronic Design*, 26(14):40-44, 1978.

14. Bursky, D. An overview of programmable logic and memory devices. *Proc. IEEE WESCON*, part 7/0, 1979.
15. Bursky, D. UVEPROM's and EEPROM's crash speed and density limits. *Electronic Design*, 28(24):55-66, 1980.
16. Cayton, B. Designing with nitride-type EAROM's. *Electronics*, 50(19):107-113, 1977.
17. Fukushima, T. The bipolar PROM: an overview and history. *Proc. IEEE WESCON*, part 7/1, 1979.
18. McMullen, J. Programming the PROM: a make or buy decision. *EDN*, 19(5):59-62, 1974.
19. Stark, M. Two bits per cell ROM. *IEEE COMPCON*, 209-221, Spring 1981.
20. Stewart, R. G. High denisty CMOS ROM arrays. *Proc. IEEE Journal Solid-State Circuits*, SC.12:502-506, 1977.
21. Tanimoto, M., Murota, J., Ohmori, Y., and Ieda, N. A novel MOS PROM using a highly resistive poly-Si resistor. *Trans. IEEE*, ED.27:517-520, 1980.
22. Tsantes, J. Programmable-memory choices expand design options. *EDN*, 25(1):80-98, 1980.
23. Twaddell, W. Uncomitted IC Logic. *EDN*, 25(7):89-98, 1980.
24. Twaddell, W. EEPROM gains in density and speed threaten to dispell UVEPROMS. *EDN*, 26(2):37-46, 1981.
25. Wallace, R. K. and Learn, A. J. Simple process propels bipolar PROMS to 16 K density and beyond. *Electronics*, 53(7):147-150, 1980.
26. Woods, M. H. A EPROM's integrity starts with its cell structure. *Electronics*, 53(18):132-136, 1980.
27. Wood, R. Evaluate UV EPROM data retention. *Electron. Des.*, 26(7):82-84, 1978.
28. Report. *Electrically Erasable Non-Volatile Semiconductor Memories*. (Reliability Report), Electronik Centralen, distributed by IPI, Copenhagen.
29. Orlando, R. Understanding your application in choosing NOVRAM, EPROM. *EDN*, 28(10):137-145, 1983.
30. Hurst, S. L. *Logical Processing of Digital Signals.* Crane Russak, New York, and Edward Arnold, London, 1978.
31. Larson, T. L. and Downey, C. Field programmable logic devices. *Electronic Engineering*, 52:37-54, January 1980.
32. Noach, K. A. H. New developments in integrated fuse logic. *Elec. Compon. Applic.*, 4:111-123, February 1982. (Reprinted as Mullard Technical Publication M82, 0032.)
33. Calvan, N. and Cline, R. C. Field PLA's simplify logic design. *Electronic Design*, 23(18):84-90, 1975.
34. Mitchell, T. W. Programmable logic arrays. *Electronic Design*, 19(15):95-101, 1976.

35. National Semiconductor. How to design with programmable logic arrays. National Semiconductor, Inc. Report No. AN. 89, Santa Clara, CA, 1973.
36. Texas Instruments. MOS programmable logic arrays. Texas Instruments, Inc. Application Bulletin CA.158, Dallas, TX, 1973.
37. Calvan, N., and Durham, S. J. Field-programmable arrays: powerful alternatives to random logic. *Electronics*, 52(14):109-114, 1979.
38. Calvan, N., and Cline, R. C. FPLA applications: exploring design problems and solutions. *EDN*, 21(7):63-69, 1976.
39. Bennett, S. HPL: The logical solution. Application Notes, Modules 1 to 5, Harris Corporation, Melbourne, FL, 1982.
40. Bennett, S., and Fisher, B. J. Speed optimization for the field-programmable logic array and other programmable logic architectures. Application Note 107, Harris Corporation (undated), Melbourne, FL.
41. Birkner, J. et al. *Programmable Array Logic Handbook*. Monolithic Memories, Santa Clara, CA, 1978.
42. Monolithic Memories. *Bipolar LSI Databook*. Monolithic Memories, Santa Clara, CA, 1982.
43. Bostock, G. H. New developments in Integrated fuse logic. *Proc 2nd Int. Conf. on Semicustom IC's*, section 3, London, November 1982.
44. Chidzey, P. Programming and testing of programmable logic devices. *Electronic Product Design*, 4:81-85, September 1983.
45. Muehldorf, E. I. High-speed integrated circuit characterisation and test strategy. *Solid State Technol.*, 93-99, September 1984.
46. Arndt, L. *Field programmable logic array*. *Fairchild Journal of Semiconductor Progress*, 6(4):3-10, Mountain View, CA, 1978.

4

Masterslice Semicustom Techniques

In Chapter 1 we briefly introduced the underlying reasons why custom-specific microelectronics has evolved to supplement (and indeed possibly supplant?) standard off-the-shelf integrated circuit (IC) packages. The different possible approaches to the economic reailzation of special circuits for original equipment manufacturer (OEM) use were also briefly introduced, which at present fall into two categories: (1) master-slice arrays and (2) cell-libraries. In this chapter we will consider in greater depth the first of these, with the following chapter considering the second range of possibilities.

4.1 BASIC CONCEPTS

It will be recalled from Chapter 1 that a masterslice array consists of a standard chip layout containing an array of standard cells, which is fabricated up to but excluding the final metal interconnections as a standard product for custom use. The customization procedure consists of the design of the on-chip interconnections necessary to realize a particular custom requirement, followed by the mask making and fabrication for the interconnections, plus final packaging and test.

The fundamental advantage of the masterslice approach of course is that a full set of masks does not have to be made for each custom application; only the dedication masks are required. In theory the uncommitted wafers can be stockpiled until such time as required for various applications. The main disadvantage is that the utilization of

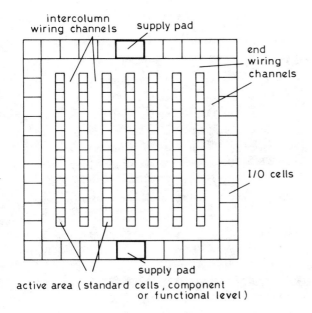

FIGURE 4.1   Typical masterslice gate-array floor plan layout.

the on-chip silicon area is not as high as alternative approaches, since almost inevitably there will be unused parts of the standard array, and performance will generally be inferior because of looser packing densities.

Figure 4.1 shows the basic concept. In the following sections we will consider in greater depth factors such as the choice of standard cell laid down on the chip, wiring channel space considerations, and further features which are involved.

4.2 UNCOMMITTED COMPONENT-CELL ARRAYS

Table 1.8 illustrated how masterslice arrays may be subdivided into the two principal categories of (1) uncommitted component-cell arrays (UCAs), in which each standard cell in the array consists of some assembly of separate individual and generally unconnected components, and (2) uncommitted functional-cell arrays (UGAs), in which each standard cell is a dedicated fully functional logic gate or other working entity. In this section we will consider the first of these two categories, although we will subsequently note that there are some masterslice arrays which combine both of these approaches.

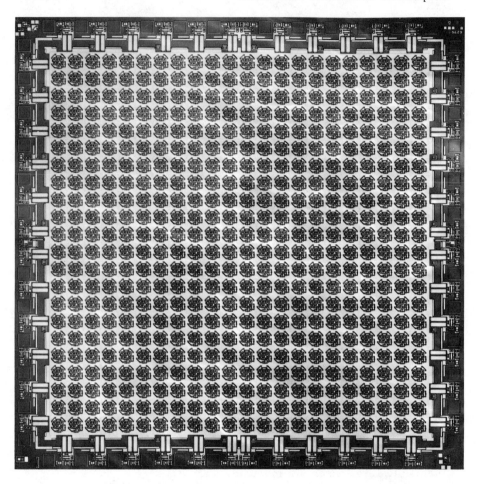

FIGURE 4.2 The Ferranti Uncommitted Logic Array masterslice topology, showing an uncommitted die and subsequent custom metalization. (Details courtesy Ferranti Electronics, Oldham, England.)

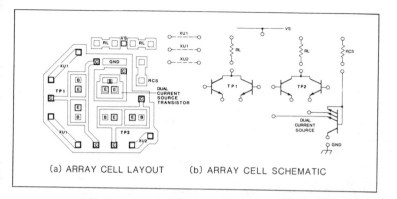

(a) ARRAY CELL LAYOUT   (b) ARRAY CELL SCHEMATIC

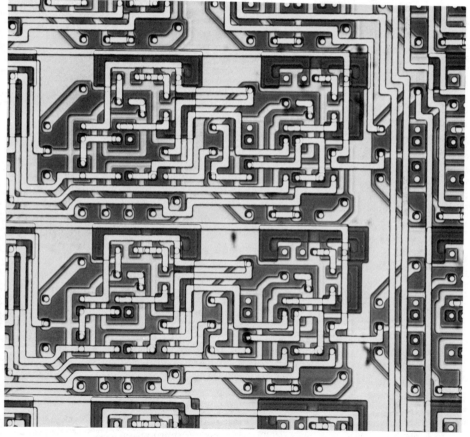

FIGURE 4.2   (Continued)

4.2.1 Bipolar UCAs

It is possibly appropriate to consider first one of the very earliest commercial masterslice products, which has stood the test of time since 1972. This is the bipolar uncommitted logic array (ULA) of Ferranti Electronics, in the United Kingdom. An early small version is illustrated in Figure 4.2. Later considerably larger versions, second-sourced by Interdesign, Inc., Scotts Valley, California, give a broad range of product availability.

Each component cell on the chip will be seen to consist of three resistors and five transistors, together with V_{CC} and ground connections. The 5 V bipolar technology is the collector-diffusion-isolation (CDI) current-mode technology (CML) covered in Chapter 2, which while not a general industry standard, has considerable advantages particularly for semicustom applications. Each cell may be internally connected so as to realize a 2- or 3-input NOR gate or 2-input NAND, or in conjunction with adjacent cells higher logic primitives and also (very importantly) analog requirements. (See also Chapter 2, Figs. 2.14 and 2.15.)

The chip architecture is in "block-cell" or "sea-of-gates" form, that is the cells are not tightly assembled in columns (or rows) as generally suggested in Figure 4.1, but instead are isolated from each other, with wiring space available around each cell. Additionally, each cell has three corner crossunders in its construction, to facilitate final chip interconnection routing. The advantage of such an open floor plan is that it provides maximum freedom for the dedication interconnections (routing), but this freedom is very difficult to exploit by computer-assisted design (CAD) software. We will cover placement and routing later in this chapter. In general this early form of masterslice layout is handrouted, employing the high ability of the human designer to do two-dimensional routing problems. The disadvantage of manual handrouting, however, is the time factor and potential errors, both of which invite the use of automatic routing, albeit not necessarily as efficient or as clever as handrouting under unrestrained circumstances.

Many later versions of this masterslice followed. As noted in chapter 2, the CML technology can be tailored to provide low power or high speed extremes, and therefore a range of system speeds is currently available, as shown in Figure 4.3, with equivalent 2-input gate counts from 450 to 10,000 gates per chip. Alternative series specifically to provide analog custom circuits are also available.

The other bipolar technologies which are represented in the semicustom area are emitter-coupled logic (ECL), low-power Schottky transistor-transistor logic (TTL), integrated-injection logic (I^2L), integrated Schottky logic (ISL), and Schottky-transistor logic (STL) (see Chapter 2 for definitions and fabrication details). ECL is for the

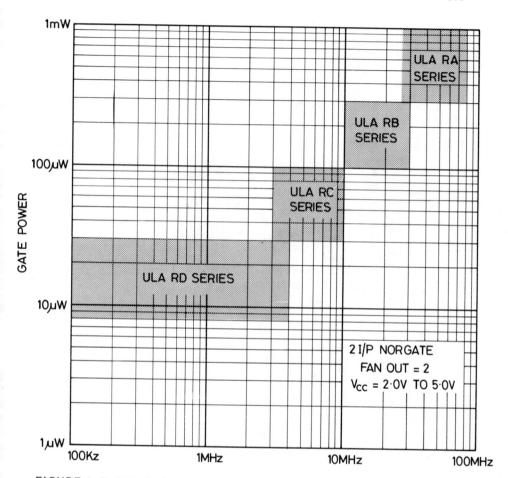

FIGURE 4.3 CML technology capability in Ferranti/Interdesign masterslice arrays.

highest speed requirements, and is limited in gate count by on-chip dissipation. Representative of current availability in masterslice component array form are the AMCC Q700 series arrays, illustrated in Figure 4.4. Two basic cell structures are provided per chip, as shown, together with fixed threshold cells to provide the ECL current levels and appropriate input/outputs (I/Os) to translate from the internal ECL working levels to the outside world requirements. A single 5 V supply is required, with on-chip translation from TTL to ECL voltage levels.

On-chip component interconnection provides the usual range of ECL gate configurations. Gate delays of less than 1 ns are internally available under minimum loading conditions, but this high speed cannot be brought off-chip via the normal TTL or ECL 10K compatible I/Os.

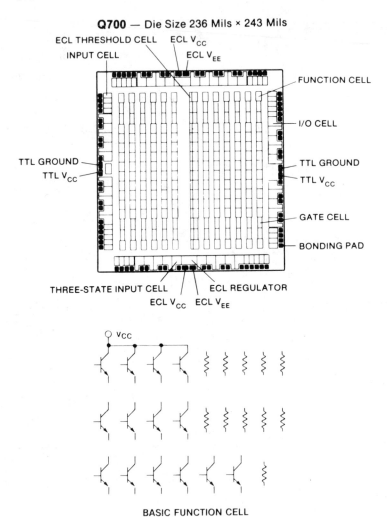

FIGURE 4.4 The AMCC Q700 ECL high-speed masterslice array.
(Details courtesy Applied Micro Circuits Corporation, San Diego, CA.)

Other ECL products have similar capabilities. The remaining four
bipolar technologies do not have strong representation in the uncom-
mitted-component array field, possibly because it is more relevant to
provide fully functional gates rather than separate components in these

technologies. One of the very few known exceptions is the Fairchild GT 0750 masterslice shown in Figure 4.5. The technology is the Fairchild advanced Shottky-TTL "FAST" technology, giving typical gate delay times of 2 ns, with a typical power dissipation per chip of 1.5 W. Assembly of the 720 cell chip with 76 I/Os is made on an 84-pin chip carrier package.

The remaining bipolar technologies of I^2L, ISL, and STL, do not appear to be commercially available except in fully functional gate-array form, as will be covered later.

4.2.2 Metal-Oxide Semiconductor UCAs

Looking at all the available metal-oxide semiconductor (MOS) technologies, and whether they are applied to the uncommitted component array field, we find a situation summarized in Table 4.1.

It is initially suprising that the present mainstream large-scale integrated (LSI) technology, namely single-channel MOS, is not strongly represented in the masterslice field, particularly in view of its well established fabrication processing. However, in nonhandcrafted situations, the required on-chip interconnection area means that single-channel MOS arrays show little if any area saving over complementary MOS (CMOS), and since the latter has lower power dissipation and generally similar speed capability for custom applications it is usually preferred.

An example of a masterslice nMOS UCA is a design used for research and development and educational purposes by Edinburgh University. The larger of the two variants has the following specification:

24 × 24 Array, 576 cells total

Die size 5.08 × 5.08 mm

V_{DD} 5 V, substrate bias V_{BB} − 2.5 V

40 Programmable I/Os

Each cell 3 enhancement-mode transistors, 1 depletion-mode transistor, plus underpasses

However, apart from such rare examples, it is the bulk silicon substrate CMOS technology, either metal-gate or increasingly silicon-gate, which dominates the commercial marketplace.

The components of CMOS are of course merely the p- and n-channel enhancement-mode active devices, and since in conventional CMOS logic

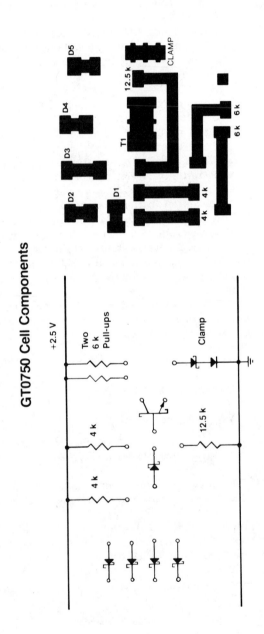

GT0750 Cell Components

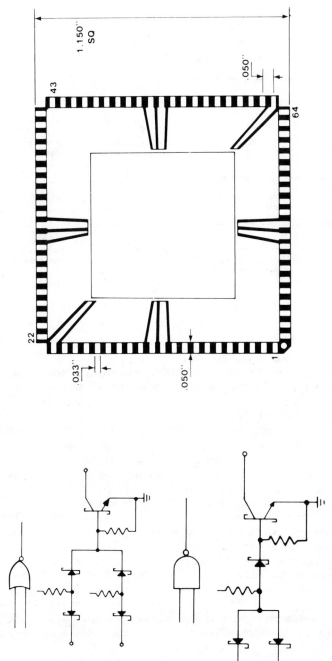

FIGURE 4.5   The Fairchild Schottky-TTL masterslice array components per cell.   (Details courtesy Fairchild Camera and Instrument Corporation, Milpitas, CA.)

TABLE 4.1   The Hierarchy of MOS Technology and Its Application to Masterslice Uncommitted-Component-Arrays

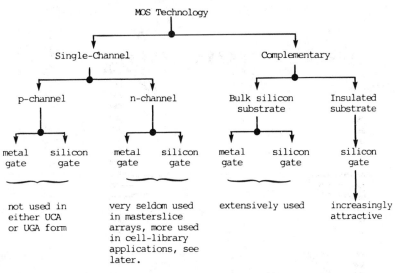

they are always used in pairs, a regular cell structure of transistor pairs readily follows. The overall chip architecture generally follows that of Figure 4.1, consisting of columns (or rows) of active elements interspersed with wiring channels, perimeter wiring channels and I/Os completing the chip layout.

The arrangement of p- and n-channel transistor pairs is generally as shown in Figure 4.6a. The transistors may be entirely separate on the uncommitted chip, or grouped as shown in Figure 4.6. If the latter is the case then one cell contains 2 p-channel and 2 n-channel devices, which gives a 2-input NAND or NOR capability per cell; if the former, then the concept of an individual cell tends to dissolve, and a complete column in the array may consist of a large number of p- and n-channel transistors stepped and repeated at equal intervals, possibly with underpass facilities at appropriate intervals.

V_{DD} and V_{SS} supply rails run along the device columns, V_{DD} being adjacent to the p-channel devices, V_{SS} adjacent to the n-channel devices. Polysilicon underpasses are normally provided to give both-side access to gates and possibly the source and drain connections, and perhaps to provide straight-through connections from one wiring channel to an adjacent channel. Figure 4.6b indicates how a cell may be committed to form a 2-input NAND gate.

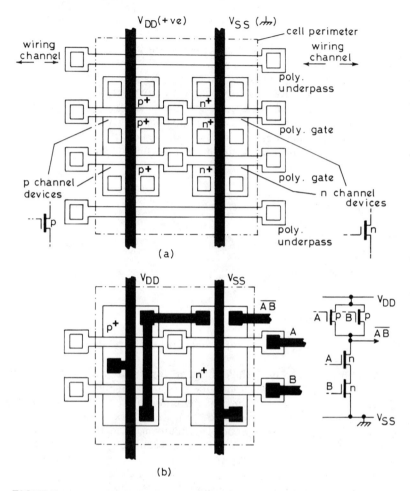

FIGURE 4.6 CMOS masterslice components per cell: (a) the general topology of *p*- and *n*-channel transistors per cell (not to scale); (b) dedication of (a) to form a 2-input NAND gate.

While this represents the basic chip architecture of the very many CMOS masterslice arrays on the market, both metal-gate and silicon-gate, possibly the greatest divergence between the different commercial products is in the interconnection strategy. This involves different polysilicon-level connections, wiring channel capacities, and perimeter wiring strategies, the problem always present being that of trying to fit any required custom interconnection pattern into the fixed wiring space provided in the chip layout. We will refer to wiring dedication later in this chapter.

Insulated-substrate CMOS is also represented in this area by silicon-on-sapphire (SOS) devices, although in relatively small numbers due to bulk silicon CMOS being cheaper and generally adequate for the custom-specific market. RCA is a prominent source of SOS-CMOS semicustom at present, with their gate universal array (GUA) product range for military and special commercial purposes.* The specification of their GUA TCS.093 masterslice, whose cell structure is as in Figure 4.6, and which is dedicated by a single metalization mask stage, is as follows:

12 × 48 Array, 576 cells total

Die size 6.1 × 6.1 mm

V_{DD} 3 – 12 V

64 I/Os

Maximum clock speed 40 MHz

Each cell 2 p-channel and 2 n-channel enhancement devices

The only known radically different CMOS layout to that shown above is that of the Racal silicon-gate CMOS cell, shown in Figure 4.7. The cell consists of four pairs of p- and n-channel transistors, arranged in a beautifully symmetrical "snowflake" pattern. Six underpasses are also provided per cell. Interconnection wiring may be taken from any side of any cell. A similarity of this design philosophy and that of the Ferranti ULA illustrated in Figure 4.2 may be noted; both provide very flexible routing facilities, but it is this very flexibility which makes it difficult to apply CAD placement and routing.

4.2.3 Further Masterslice Products

There are further masterslice products on the commercial market which do not fall precisely into the previous well defined categories of regular

---

. *The jungle of terminology in the custom and semicustom area is frequently troublesome. It is sometimes difficult to be sure from commercial literature whether a device requires only interconnect dedication mask(s) for its commitment or a full mask set, and whether or not dedication interconnections at component level have to be undertaken.

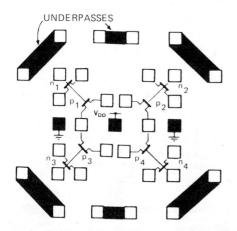

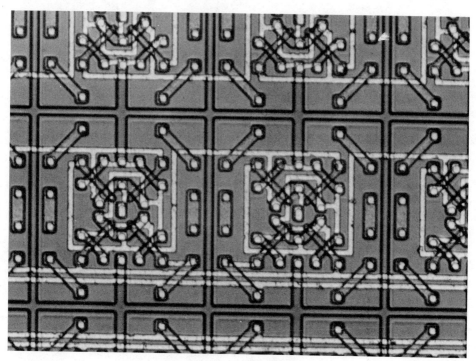

FIGURE 4.7 The Racal "snowflake" CMOS cell used in masterslice arrays. (Details courtesy Racal Microelectronic Systems Ltd., Reading, England.)

structure arrays, either because their chip layout is not a regular matrix of identical cells, or because they contain an admix of both component cells and fully functional cells. Among such further variants may be found:

1. Uncommitted die containing a wide selection of various components
2. Uncommitted component-cell arrays with an additional selection of various components per die
3. Uncommitted component-cell arrays with some further fully functional cells per die
4. Uncommitted die containing a selection of various fully functional primitives

As examples of (1) above, the component specification for two commercial products are given below:

Interdesign Inc., Monochip type MOF

Linear bipolar

437 Components, die size 2.31 × 2.79 mm

24 I/Os

24 V max.

92 Small *npn* transistors

4 Power *npn* transistors

36 Dual *pnp* transistors

18 200 Ω resistors

88 450 Ω resistors

68 900 Ω resistors

61 1.8 KΩ resistors

61 3.6 KΩ resistors

9 30 KΩ resistors

Cherry Semiconductor, Genesis 3500

Linear bipolar

299 Components, die size 1.91 × 2.46 mm

22 I/Os

20 V max.

59 *npn* Transistors

24 *pnp* Transistors

2 Power transistors

214 Resistors (various)

1 Photodetector

An example of the second form of hybrid specification may be found in the Master-Chip XR 400 product of Exar Integrated Circuits Systems. This combines a fast I^2L logic array with additional linear components, so as to provide moderately complex analog and digital capability. The general specification is:

Die size 3.02 × 3.96 mm

40 I/Os

7 V max.

256 Cell gate array, each cell one 5 output I^2L logic gate

45 *npn* Transistors

12 4-Collector *pnp* transistors

200 700 Ω resistors

116 2.5 KΩ resistors

20 5 KΩ resistors

An illustration of this product is given in Figure 4.8.

The third variant, consisting of an uncommitted component-cell array with additional fully functional cells, may be found in the American Micro Circuits Corporation range of CMOS products, such as the midrange Q413 device. This contains a total of 72 component-level cells, each of which consists of 10 *p*-channel and 10 *n*-channel devices, together with 36 dedicated static D-type bistable circuits. Additionally, there is provided on-chip clock generation, together with up to 52 I/O pads for I/O duties. Maximum clock frequency is 3 MHz, and typical power dissipation at 5 V and 80% chip utilization is 225 mW.

Finally, one may note the availability of various fully functional primitives per die, as instanced by WD 1820 Logic Array Device of Western Digital. This particular product contains:

37 Inverters

10 Noninverting buffers

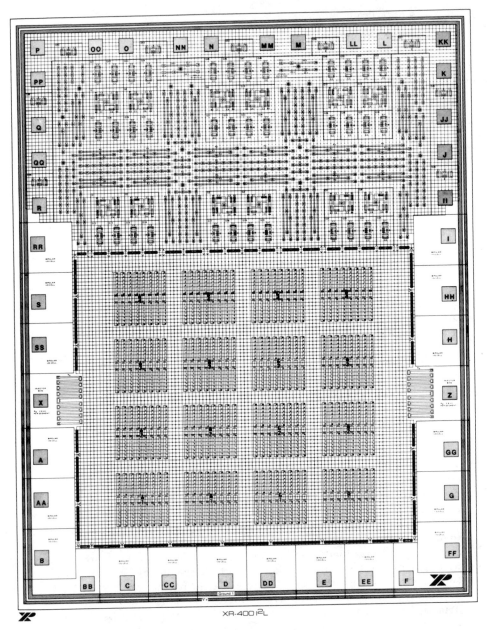

FIGURE 4.8   The Exar XR 400 Master-Chip masterslice gate plus discrete components chip.   (Details courtesy Exar Integrated Systems, Inc., Sunnyvale, CA.)

38 2-Input NOR gates

19 2-Input NAND gates

4 2-Input EXCLUSIVE-OR gates

12 D-type bistables

10 JK-type bistables

16 Programmable I/Os

2 Input-only inputs

The on-chip circuits are bipolar, fully compatible with 74 LS-series small-scale and medium-scale integration (SSI and MSI) packages. Packaging is 20-pin dual-in-line (DIL), (see Fig. 4.34).

It will be apparent that the majority of these forms of masterslice wafers are most applicable for hybrid applications, rather than digital only requirements. They particularly fill a place for special applications, especially where a range of current and voltage applications may be required. Due to the distributed layout of components in the majority of these circuits, customization is by hand routing. Single-layer metal is usual. CAD resources may be employed for circuit simulation and checking, as well as for the preparation of the final interconnect design fabrication.

For futher details of all the types of masterslice products covered in the preceding sections, reference may be made to suppliers' data sheets (see also Section 4.8), and to many other sources [1—23].

4.3 UNCOMMITTED FUNCTIONAL-CELL ARRAYS

Turning now from uncommitted component arrays to masterslice products which contain an array of fully functional elements (primitives) we find that the range of technologies previously considered are also represented in this area. The initial attraction of primitives instead of components is the elimination of the need for any internal cell interconnection of the components, thus saving on design time, interconnect details, and silicon area, but a disadvantage is a loss of flexibility and universality.

The immediate question which arises as soon as masterslice UGAs are considered is what shall be the specification of the primitives(s) chosen to be the standard cell in the array? Clearly the choice must be a circuit from which all system requirements can be made. We will

first consider this question in some detail before continuing to show
current commercial examples.

Should we be dealing entirely with analog requirements, then it is
possible to suggest an array of cells, each of which consists of, say,
a differential-input operational amplifier. This could hardly be called
a gate array and the terminology "uncommitted amplifier array," would
be appropriate.

As far as is known, no such commercial product exists. The majority of custom requirements involving analog duties are much more likely
to be hybrid rather than entirely analog, and hence uncommitted products such as described in the previous section are appropriate. One
further commercial product range may be noted; this is the CMOS technology Semicustom Analog/Digital Component Array type HI.3700 of
Holt Integrated Circuits, Inc., which, in spite of being termed a "component array" contains dedicated elements, including 11 operational
amplifiers, 22 D-type bistable circuits, 1 8-bit digital-to-analog con-
verter, 1 comparitor, and other items.

4.3.1 The Choice of Standard Logic Primitive

The logic primitive chosen as a standard cell must be capable of being
used as a building block from which all possible digital requirements
can be assembled. It must therefore be a universal logic element.

The simplest choice of logic primitive is either a NAND or NOR gate;
both are universal logic elements, and are widely chosen for masterslice gate arrays. However, such a choice may be inefficient due to the
limited logical discrimination of such gates; by themselves each gate
can only uniquely distinguish one out of the 2^n different possible binary input patterns which may be applied to the gate inputs, where
n is the gate fan-in, and therefore an appreciable number of such
gates require to be used to realize higher logical requirements.

In addition to simple NAND or NOR gates, a number of more comprehensive logic configurations may be proposed which are functionally
universal. In particular, the more comprehensive ones are characterized by the feature that their inputs are not symmetrical, that is they
realize different output functions depending upon the pattern of their
input connections, including logic 0 and logic 1 as inputs, which is
unlike simple NAND, NOR, and other Boolean gates which always realize the same logical relationship irrespective of input connection permutations. The specification of such primitives, otherwise known as
universal logic modules (ULMs), may be derived from function classification theory. Some of the possibilities are shown in Figure 4.9.
Possibly the best known is the multiplexer, but there are many candidates, some of which show greater logical flexibility than others, although by definition all are universal [24-27].

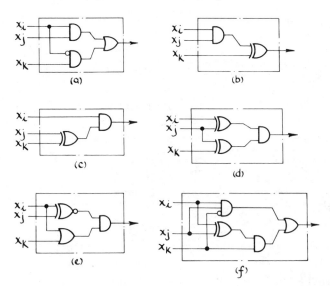

FIGURE 4.9   Some simple single-output universal logic modules revealed by function classification theory. (NAND and NOR gates are additional to these examples.)

For practical purposes we are interested that the number of input connections per cell shall be a minimum for a given range of universality [27,28]. However, the final selection of a primitive for general-purpose use should not only take into account its general specification, but must also include consideration of its efficiency in terms of the silicon area occupied both for the layout of the cell itself and for the number of dedication interconnections between cells. An optimum choice would be one which

1. Possesses high logical capability, thus reducing both the number of cells and the number of interconnections necessary to realize a given system requirement
2. Possesses the ability to realize efficiently sequential functions
3. Can be laid out in a small silicon area, ideally size-limited by the number of input and output connections per cell rather than internal circuit complexity

These interrelated factors are contrasted in Table 4.2 below.

It is perhaps clear that there is no one optimum solution in this problem. NAND/NOR gates currently represent the most widespread choice in commercial masterslice arrays, but this is principally due to familiarity rather than any serious technical evaluation of possible

TABLE 4.2   The Decision Factors in Choice of Standard
Masterslice Cell

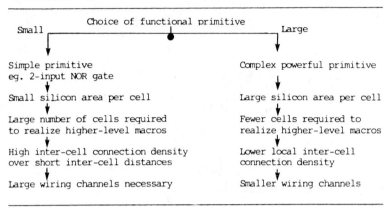

alternatives. However, there have been in-depth studies of alternative possibilities, from which it is indicated that higher power primitives than NAND/NOR are optimal. These studies indicate that a particular four-input, two-output primitive has attractions, although
other variants have similar advantages over the very basic NAND/NOR
choice. Additional details of these studies may be found [25–31]; we
will refer to this further in Sec. 4.4. However, there is a strong
streak of conservatism in basic logic design in spite of its great sophistication at CAD and fabrication level, and any changes from accepted
Boolean primitives are slow. Thus, in the following sections dealing
with commercially available products little innovation at primitive level
will be noted.

4.3.2 Bipolar UGAs

Bipolar masterslice gate arrays are available in ECL, low-power Schottky
TTL, I^2L, ISL, and STL technology. The one bipolar family not represented is current-mode CDI, which is confined to the component
array products covered in Sec. 4.2.1.
     An example of a very high-speed ECL masterslice UGA is the Fair-
child GE.0020 array. The standard cell is a large six-input, four-
output OR/NOR circuit shown in Figure 4.10a; the full array specification is as follows:

4 × 4 Array, 16 cells total, each cell 6-input OR/NOR

Die size 6.3 × 6.3 mm

V_{CC} 5 V

V_{EE} − 5 V

V_{CS} 2.5 V

24 Logic I/O's

0.3 ns Internal gate delay

0.5 W Typical chip dissipation

Larger ECL arrays generally tend to be at component level, such as previously illustrated in Figure 4.4. Power dissipation per chip is then determined by the particular utilization of the available components.

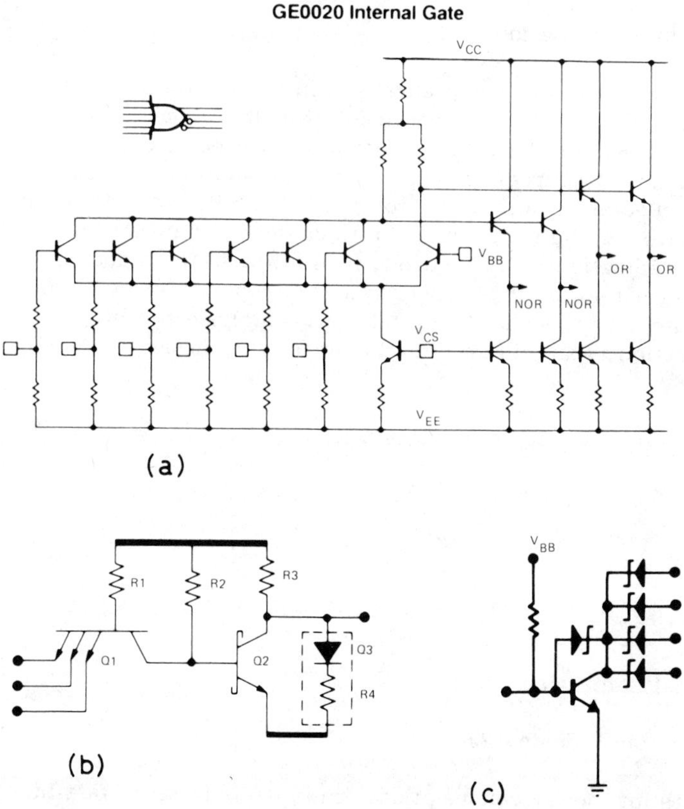

FIGURE 4.10 The logic gates employed in various masterslice gate arrays: (a) high-speed ECL OR/NOR gate of Fairchild GE.0020 array; (b) low-power Schottky TTL gate; (c) high-speed STL gate.

Low-power Schottky TTL is represented in many commercial gate-arrays. The majority have 3- or 4-input gates, such as illustrated in Figure 4.10b. The general specification for the Fujitsu B.500 master-slice array using this cell is:

8 × 8 Macro cells, each macro eight individual basic cells, 512 cells total, each cell 3-input NAND

Die size 4.9 × 5.1 mm

V_{CC} 5 V

60 Logic I/Os

1.8 ns Internal gate delay

1 W typical chip dissipation

I^2L technology has fewer representatives in the masterslice gate-array field than Schottky TTL. The basic I^2L gate is similar to that shown in Figure 2.8a, but usually has certain features in the chip layout in that current sources (injectors) may be shared between more than one multicollector output transistor, and flexibility of gate input and output points may be available. Three custom masks may be in-volved in chip dedication [32]. Indeed, it is interesting to decide whether an I^2L masterslice is an uncommitted component array or an uncommitted gate array, since it may fall between these two stools. A typical specification is as follows, this being for the Siemens S.360B masterslice array:

1584 Cells total, each cell a single-input, three-output gate

Die size 6 × 5 mm (approx.)

V_{CC} 5 V

45 Logic I/Os

15 ns Typical gate delay

0.5 W typical chip dissipation

ISL, with its low power consumption, lends itself to large fast bi-polar masterslice arrays, such as that illustrated in Figure 4.11. In most ISL arrays, the floor plan layout of the ISL cells is augmented

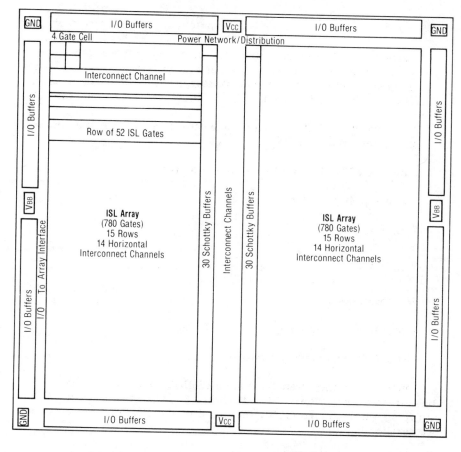

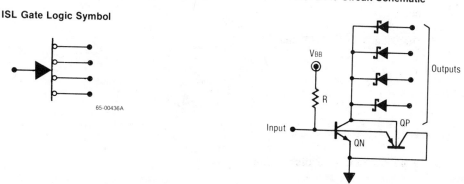

ISL Gate Logic Symbol

65-00436A

ISL Gate Circuit Schematic

FIGURE 4.11   The general floor plan architecture of the Raytheon CGA.1800 ISL masterslice UGA and its ISL standard cell. (Details courtesy Raytheon, Mountain View, CA.)

191

with Schottky buffer cells to provided increased internal fan-out capability, together with the necessary peripheral I/Os to interface between the outside world and the on-chip voltage levels [33]. The general specification for the Raytheon chip of Figure 4.11 is as follows:

2 Arrays of 52 × 15 ISL cells, 1560 cells total, each cell a single-input, four-output gate

2 Arrays of 30 Shottky buffer cells, 60 total

Die size 5.46 × 5.46 mm

V_{CC} 7 V

V_{BB} 4 V

76 Logic I/Os

3 ns Internal gate delay

250 mW Typical chip dissipation

Finally, Schottky-transistor logic (STL). One masterslice example has already been illustrated in Figure 1.5. The standard cell of this array shown in Figure 4.10c, the general specification for the whole masterslice being:

26 × 30 Array, 1560 cells total, each cell a single-input, four-output cell

Die size 5.68 × 5.68 mm

V_{CC} 5 V

108 Logic I/Os

2 ns Internal gate delay

0.45 mW per gate dissipation maximum

In summarizing this bipolar area, it will be seen that the emphasis is toward high speed. A disadvantage with a number of families is the initial discomfort of having to design using single-input, multioutput logic gates, rather than more familiar multiinput, single-output primitives. The MOS/CMOS arrays which we will now consider do not involve this feature, and hence, if absolute speed is not required tend to be more user-friendly than bipolar.

4.3.3 MOS UGAs

Referring back to Table 4.1, it is only *n*MOS and CMOS technology which we need consider; *p*MOS is not a representative technology in the masterslice field.

*n*MOS technology is still not well represented. There are a few masterslice uncommitted gate-arrays available, among them being the uncommitted functional array of ITT Semiconductors (United Kingdom), types LAN 1200 and LAN 2000. The standard cell specification however is unusual in that it does not consist of one single logic primitive, but rather a small assembly consisting of the following separate functional items:

1 Negative edge-triggered type-D bistable circuit, with direct set and clear (preset and reset)

1 2-Input NAND gate

6 2-Input NOR gates

1 INVERTER gate

The type D circuit can be converted if required into a type JK by means of the remaining gates per cell.

The smaller LAN 1200 product contains 70 such cells with 26 logic I/Os, while the larger LAN 2000 contains 120 cells with 34 I/Os. Supply voltage is 5 V, I/Os are fully TTL-compatible, and maximum clock speed is 3 MHz. Single-layer metalization dedication is employed, the cells being arranged in pairs in a sea-of-gates architecture, with wiring channels on all four sides of each cell pair .

A further *n*MOS gate-array which contains more than one primitive is the universal logic gate array, (ULGA) (yet another designation) of the Royal Signals and Radar Establishment (United Kingdom). This chip was designed to allow "consideration of logic design at the gate level rather than at the transistor level" with the merit that "reduced intracell metalisation enhances intercell connection transparancy." The general architecture is shown in Figure 4.12; a design speed of 50 MHz maximum is quoted, with maximum power dissipation per chip of 1/2 W at 80% cell utilization. Scaling of this design to smaller device geometries is forecast.

Both UGAs and other commercially available *n*MOS arrays often give an impression of being interim products which will be superceded by

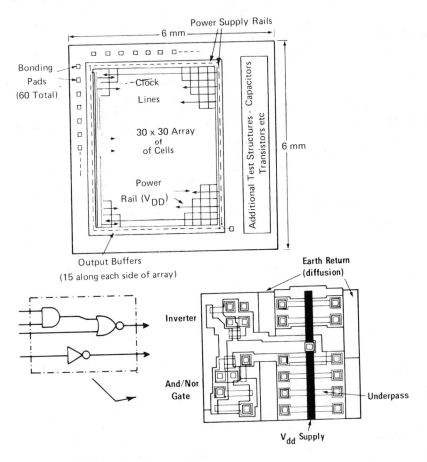

FIGURE 4.12   The RSRE ULGA masterslice uncommitted gate array in a MOS technology. (Details courtesy RSRE, Malvern, England.)

CMOS equivalents in due course. Whether this will be so remains to be seen, particularly since nMOS theoretically is a somewhat more simple fabrication process and marginally a higher speed technology.

Turning therefore to CMOS, we find that the balanced structure of equal numbers of p- and n-channel devices per conventional logic gate has to date largely led suppliers to design transistor-pair masterslice arrays, as shown in Figure 4.6. This in spite of the fact that CMOS lends itself particularly well to the design of more complex logic configurations by appropriate simple series and parallel connections of the p-channel devices and corresponding parallel and series connections of the n-channel devices. Very elegant, compact, and powerful primitives can be handrafted in CMOS, but since chip specification and design is largely in the hands of IC manufacturers rather than the logic designers who eventually wish to use the product, conservatism at cell level is prevalent. Complex CMOS gates such as shown in Figure 4.13, are certainly built up by logic designers in using CMOS masterslice transistor-pair arrays and in handcrafted CMOS libraries (see Chapter 5), but not at dedicated standard-cell level.

Hence, we have an anomalous situation at present, in that although CMOS is widely regarded as the most promising technology for the custom and semicustom field, there is no widely known CMOS masterslice array which employs dedicated logic primitives in its design. Instead we find 1-pair cells, 2-pair cells, 3-pair cells, and 4-pair cells, with the general topology illustrated in Figure 4.6 as the representative present commercial status [1].

At the research and development stage, however, we may cite some recent developments. Recalling Sec. 4.3.1 and Figure 4.9, there are a number of logic primitives other than Boolean NAND/NOR gates which may be candidates for adoption as standard function cells. The simple multiplexer cell has been used in some commercial applications as a logic primitive (not in CMOS masterslice gate arrays), but not as far as is known any alternative possibilities. However, recent work has suggested that the relatively complex primitive illustrated in Figure 4.14a may be a powerful masterslice candidate. This circuit will be seen to be an enhanced form of multiplexer, input y_1 multiplexing either input y_2 or the product term $\bar{y}_3 y_4$ to the output z_1. A complementary output z_2 is also provided.

The particular attributes of this candidate are:

1. All 2-input functions, many 3-input functions, and some 4-input functions are available from one cell.
2. Latch and type D bistable functions are available per cell.
3. Very compact silicon layout in CMOS technology is available.

The logical power of the cell is equivalent to about three 2-input NAND gates. Note that it is the pattern of the logic input signals, including logic 0 and 1, which determine the cell function; the cell therefore is a hard-wired programmable form of circuit, with no variable internal cell connection programming being necessary.

A CMOS layout for this cell is shown in Figure 4.14b. Because all dedication connections are made to the cell terminals, the internal cell layout can be compacted to within the limits of the chosen technology, no internal cell space having to be left for contact vias and for metalization. Both cell outputs are fully buffered to provide good on-chip driving ability.

Further details of this cell and the CMOS chip on which it has been fabricated may be found elsewhere [29−31]. Although at first

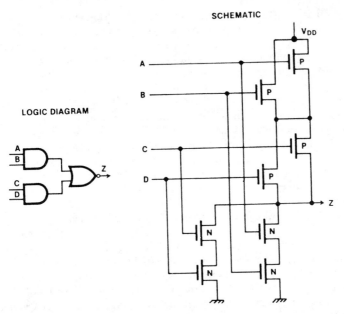

FIGURE 4.13   Possible complex logic primitives readily laid out in CMOS technology.

encounter it may appear to be an unorthodox and difficult primitive to use as a standard digital building block, precise details of its internal circuitry are completely unnecessary for its use; instead all that is required is a readily prepared look-up table of required cell I/O connections to realize a range of standard logic macros, which is no more difficult than looking up the pin connections on standard off-the-shelf SSI and MSI logic packages. Indeed, the parallel between these design procedures is extremely close.

This and similar universal logic module primitives would appear to be very appropriate for the general-purpose semicustom field. It relieves the designer from the need to descend to transistor level during his dedication design, and mirrors more closely conventional non-semicustom system design. The use of higher functional level primitives also reduces the dedicated cell interconnection density on-chip, thus reducing both custom design time and also the wiring channel capacity which needs to be provided on the masterslice layout.

It is inevitable that with any higher functional level primitive that it will occasionally be underutilized. For example, it would be a gross underutilization to use the cell of Figure 4.14a as a simple

INVERTER gate, but due to its universality this would seldom if ever be required.* Thus there will always be a debate on what constitutes an optimal choice. The one illustrated here is not necessarily the "best," and therefore scope for individuality still remains. However, the compact handcrafted nature of a functional cell and the reduction of interconnections means that overall no great silicon area cost penalty is involved even with some underutilization of cell power.

Our various discussions on these aspects are finally summarized in Table 4.3 following. See also the listed references for additional data [1−23, 32−34].

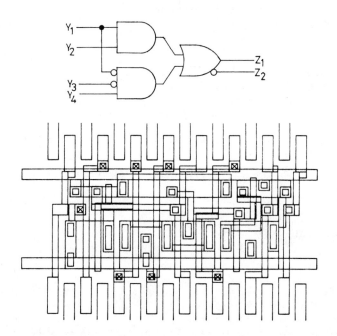

FIGURE 4.14 A universal logic module (ULM) primitive appropriate for masterslice UGAs: (a) the ULM primitive; note not fabricated in this precise AND/OR form; (b) CMOS layout of the primitive.

---

*The primitive of Figure 4.14 can equally realize the functions AB, $\bar{A}\,\bar{B}$, $A + B$, and $\bar{A} + \bar{B}$ without the complement of either A or B being required as inputs. Considerable scope, therefore, exists for eliminating individual signal complements.

TABLE 4.3   General Commercial Masterslice Availability[a]

	Technology	Array structure: component or functional cells	
		Uncommitted component array (UCA)	Uncommitted gate array (UGA)
Bipolar	CDI	√	—
	ECL	√	√
	Low-power Schottky TTL	√ (Uncommon)	√
	I^2L	—	√
	ISL	—	√
	STL	—	√
MOS	p-Channel	(Never used)	(Never used)
	n-Channel Metal-gate	(Rarely used)	(Rarely used)
	n-Channel Silicon gate	(Rarely used)	(Rarely used)
CMOS	Metal-gate, silicon substrate	√	—
	Silicon-gate, silicon substrate	√	(In future?)
	Silicon-gate, insulated substrate	√	(In future?)

[a]Hybrid masterslice are also available; an increasing range of CMOS devices likely.

4.4 WIRING CHANNEL AND ROUTING CONSIDERATIONS

In parallel with the choice of standard cell for a masterslice array is how the floor plan layout shall be arranged. This is largely technology-independent, but is a major factor of significance since once a decision has been reached it is fixed for all dedications of the array to specific custom requirements. The provision of inadequate interconnection channels will mean difficulty (or impossibility) in routing required systems on the chip; provision for too many interconnections means waste of silicon area for all small applications for which the chip is dedicated, although this may not be a serious disadvantage. It will be appreciated that this problem does not arise in cell-library designs (see Chapter 5), since the wiring areas are tailored to fit the precise interconnection requirements per dedication.

The number of interconnections required between primitives or higher level macros in a digital system perforce varies with the particular system requirements. More generally, digital systems may be divided into those with a strong functional topology, exemplified by shift register configurations, and those with more random logic requirements. The interconnection density of the latter will be higher than that of the former. However, there have been studies of the average interconnection density present in digital logic networks, based upon the number of I/O terminals per cell or primitive, which studies have given rise to an estimate often known as Rent's rule.

Consider the division of a complete system into a number of subdivisions D, each of which consists of a number of primitives (gates or cells). If there are B primitives per subdivision, then a complete system S consists of B × D primitives. Further, let each primitive have an average of K input/output terminals. This is illustrated in Figure 4.15. Landman and Russo have analysed such a situation applied to a number of real-life digital systems [35], and by altering the number of subdivisions ranging from D = 1, B = S (= all system primitives contained within the one boundary) to the other extreme of D = S, B = 1 (= only one primitive per subdivision), relationships between the required I/O terminals per subdivisions versus the number of subdivisions per system have been made.

Provided the number of subdivisions D of the system is greater than, say, 5, then the relationship

$$P = KB^r \tag{4.1}$$

has been shown, where

K = number of terminals per primitive

P = number of I/O connections around the perimeter of the subdivision and where index r has the range

$$0.57 \leqslant r \leqslant 0.75 \tag{4.2}$$

depending upon the logic structure of the digital system. This relationship has been termed Rent's rule, although Rent (IBM, 1960) does not
appear to have personally published this result. The index r is sometimes termed Rent's index. Note that in the extreme case of only one
primitive per subdivision, we have

$$P = K(1.0)^r = K$$

which is obvious.

For the division of the complete system into a smaller number of
subdivisions, $D < 5$, then the relationship

$$P = [KB - K_b B^{r_b} - K_s B^{r_s}] \tag{4.3}$$

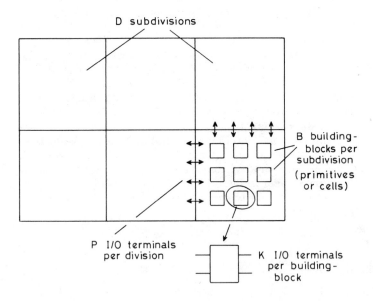

D subdivisions

B building-
blocks per
subdivision
(primitives
or cells)

P I/O terminals
per division

K I/O terminals
per building-
block

FIGURE 4.15 The division of a digital system into D subdivisions,
with B primitives per subdivision.

is suggested, where

KB = total terminal requirements if every terminal K per primitive subdivision was individually connected to an I/O terminal of the subdivision

$K_b B^{r_b}$ = an empirical saving due to burying of terminal nets entirely within a subdivision

$K_s B^{r_s}$ = an empirical saving due to sharing of nets and terminals

Unfortunately these latter two saving factors appear to have no easy value or formulation. Further, the precise boundary between region I where Eq. (4.1) holds, and region II where Eq. (4.3) holds is imprecise; the value of D = 5 used above may vary from 3 to 12 in the examples considered by Landman and Russo, with D = 5 as an average boundary value. Therefore, without precise data it would seem appropriate to ignore region II, and use when appropriate Eq. (4.1), which is what is normally considered to be Rent's rule.

A further study by Heller et al. [36] covers a statistical exercise in wiring space requirements. With the previous work these constitute the only known publications which specifically address the number of connections and/or interconnections required in digital logic networks.

Heller et al. [36] consider the wiring requirements for a number of cells, both for the one-dimensional wiring case, where all the connections are constrained to running in one wiring channel alongside rows (columns) of cells, and also for the two-dimensional case, where connections run both horizontally and vertically. Here we will only consider the one-dimensional case, since this corresponds to the main wiring requirements of most masterslice arrays (see Fig. 4.1). The first result given is that the probability that k connecting wires lie in the wiring channel immediately adjacent to any given cell in an infinitely long string of cells is

$$P_k = \frac{e^{-y\overline{R}}(y\overline{R})^k}{k!} \times 100 \qquad (4.4)$$

where

P_k = the probability of k adjacent connections

y = average number of terminals per cell (= K in our previous equations)

$\overline{R}$ = average length of the interconnections, in cell spans

Further developments give the probability P_s of successful wiring a row of R cells with a maximum wiring channel capacity of T tracks, namely:

$$P_s = \{P_0 \cdot {}_{\Pi_0}R_0{}^{-1} \cdot Q_0\} \qquad\qquad (4.5)$$

where

P_0 is a (T + 1) column vector

${}_{\Pi_0}R_0{}^{-1}$ is a (T + 1) × (T + 1) square matrix

Q_0 is a (T + 1) column vector

the elements of these vectors and matrix being based upon developments from Eq. (4.4).

Heller et al. continue their developments with a series of tabulations and graphs relating the theoretical values of track capacity versus various parameters. However, it is not clear how to obtain guidance on required track wiring capacity for general-purpose masterslice arrays, without knowledge of or assuming a number of parameter values. For further details, see published material [36−40].

Let us, therefore, here revert back to Landman and Russo and Rent's rule, applying them more directly to masterslice arrays with rows (columns) of primitives (cells).

The division of a given system in D subdivisions, as shown in Figure 4.15 can equally be drawn as a subdivision of a string of cells or primitives which consititute the complete system, into D blocks each of B primitives. This is shown in Figure 4.16a. Rent's rule now gives the number of I/O connections from each such block of B primitives, assuming as before some even distribution of interconnection density throughout the complete system.

We may redraw Figure 4.16a into the form shown in Figure 4.16b, from which the I/O connections of the B cells become the wiring channel end connections. However, since we are not assuming any local peak interconnections density, and that the I/O connection count P will theoretically be the same wherever we window our boundary in Figure 4.16a, then this value P will be twice the wiring channel interconnection density at any point along the wiring channel for the row of B cells.

As an example, suppose we consider a 400-cell masterslice arranged in a 20 × 20 array, each cell having 4 I/O terminals. Then the total number of connections entering and leaving a row of 20 cells is given by

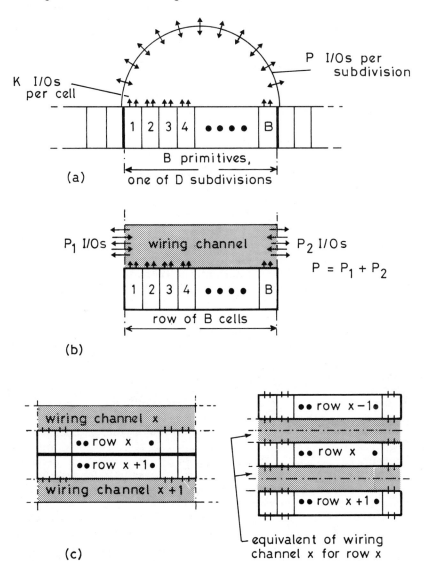

FIGURE 4.16   Revision of Figure 4.15 into a string of primitives: (a) B primitives or cells in each of D subdivisions; (b) redraw of (a) into masterslice array row and wiring channel form; (c) other masterslice row and wiring channel topologies, showing back-to-back and both-sides access variants.

$$P = KB^r$$

$$= 4\,(20)^{0.57}$$

$$= 22 \text{ taking } r = 0.57$$

$$= 4\,(20)^{0.75}$$

$$= 38 \text{ taking } r = 0.75$$

Hence, the estimated wiring channel capacity requires to be

11 Tracks for $r = 0.57$

19 Tracks for $r = 0.75$

The same calculation for differing number of cells per row, each cell still with 4 I/Os, gives the values shown in Table 4.4.

The results given in Table 4.4 are independent of the total number of cells per chip, provided the number of rows of cells is greater than 5 [the system subdivision factor noted in the development of Eq. (4.1)]. It is also independent of the actual arrangement of cells, whether back-to-back rows or both-side accessible (see Fig. 16c), but does assume that all connections have to run along single wiring channels and that two-dimensional interconnect is not present.

The validity of the results obtained from Rent's rule may be debated. What value of the index r to use is of critical significance   A study of commercial gate arrays which have a simple interconnection topology without excessive "escape and jog" tracks through cells and across wiring channels seems to indicate a provision of 10 to 14 tracks per wiring channel for about 20 functional cells per row (column). This implies that the lower estimate of Rent's rule is realistic for most master-slice arrays, accepting the risk that placement may become critical in order to eliminate possible routing bottlenecks.

The majority of the published information in this area assumes that the functional primitives of the system are conventional Boolean gates, typically NAND or NOR. However, we have seen in Figure 4.14 the suggestion of a fairly complex primitive for gate-array use, which may be hard-wired to perform a range of duties and which reduces the number of interconnections required between such primitives. Studies on the channel interconnection density using such primitives indicate a relatively low value for Rent's index; figures ranging from $r = 0.42$ minimum to 0.54 maximum have been quoted, which are both below the previously cited minimum value [39,40]. Hence, there is some indication that the choice of standard cells of higher logic capability reduces the wiring channel requirements, as should be expected.

TABLE 4.4   Rent's Rule Applied to Estimate the Number
of Wiring Tracks per Row of Cells, Each Cell with 4 I/O
Terminals (K = 4)

No. of cells per row (or column) B :	10	16	20	40	80
$P = KB^r$, r = 0.57	15	19	22	32	48
$P = KB^r$, r = 0.75	22	32	38	64	106
Theoretical wiring channel requirements					
minimum	8	10	11	16	24
maximum	11	16	19	32	53

The above considerations are mainly concerned with a linear arrangement of cells, as in Figure 4.16. However, there may be many other
suggested floor plan layouts for functional cells, including:

Sea-of-cells, with wiring space on all four sides of every cell

Double-cell rows (columns)

Quadruple-cell rows (columns)

Four-square cell assemblies

Eight-cell assemblies

and other arrangements. The above possibilities are shown in Figure
4.17.

The simplest topology to route is the linear row (column) of Figure
4.17a, since all the principal interconnections are one-dimensional. It
may not, however, be the most efficient in terms of total chip area.
The most complex to route is the sea-of-cells of Figure 4.17b, since
full utilization of the flexibility afforded by the 2-dimensional wiring
space is difficult—it may produce a more compact final result than
other topologies, but at the expense of considerable iteration and re-
placement in order to achieve optimality.

We may easily generate expressions giving the silicon area for each
topology based upon a given wiring density for each layout, such as
follows.

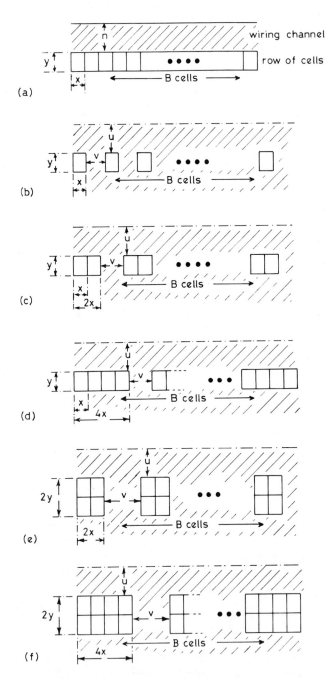

FIGURE 4.17 Possible floor plan layouts for standard cells: (a) linear rows (columns), no intercell spacing; (b) distributed sea-of-cells; (c) double-cell rows (columns); (d) quadruple-cell rows (columns); (e) four-square assemblies; (f) eight-cell assemblies.

1. Linear assembly of Figure 4.17a:

    silicon area A per B cells $= B\{x(y + n)\}$ units^2 (4.6)

    where one unit is the design pitch of the interconnections across the wiring channel

    > n = number of interconnection tracks provided in the wiring channel
    >
    > x = width of cell, in the above units
    >
    > y = height of cell, in the above units

2. Sea-of-cells, Figure 4.17b:

    $$A = B\{(x + v)(y + u)\}\ \text{units}^2 \tag{4.7}$$

    where

    > u = number of interconnection tracks in the y direction
    >
    > v = number of between-cell tracks in the x direction

3. Double-cells of Figure 4.17c:

    $$A = \tfrac{1}{2}B\{(2x + v)(y + u)\}\ \text{units}^2 \tag{4.8}$$

    where u and v are as previously but not necessarily the same value.

4. Quadruple-cells of Figure 4.17d:

    $$A = \tfrac{1}{4}B\{(4x + v)(y + u)\}\ \text{units}^2 \tag{4.9}$$

5. Four-square assembly of Figure 4.17e:

    $$A = \tfrac{1}{4}B\{(2x + v)(2y + u)\}\ \text{units}^2 \tag{4.10}$$

6. Eight-cell assembly of Figure 4.17f:

    $$A = \tfrac{1}{8}B\{(4x + v)(2y + u)\}\ \text{units}^2 \tag{4.11}$$

However, we cannot evaluate these equations to determine the minimum silicon area candidate without first allocating values for the number of wiring channels in each topology, which cannot be done by any reliable method. One can equate one equation against another for various chosen values of n, u, and v to show graphically equi-area conditions of one with respect to the others [41], but the results are still difficult to assimilate and their significance debatable.

Thus there is no scientific means of determining the "best" layout for a gate array in order to minimize the silicon floor plan area; the problem always revolves around how many interconnection channels are necessary in each topology, which in turn is largely dependent upon how efficient or easy the topology is to route.

Many different topologies may be found in the literature [41—50]. However the majority of commercial masterslice arrays employ some form of linear assembly of cells (rows or columns) with unidirectional wiring channels, as originally illustrated in Figure 4.1 and subsequently in Figure 4.17a. There are notable exceptions, for example as already seen in Figures 4.2 and 4.7, but these are individualistic approaches. Let us, therefore, consider in somewhat greater detail the problems of routing such a topology, and how the basic array floor plan may be modified in detail in the cell design or in the wiring channel design in order to improve the overall routability.

From earlier figures in Chapter 2, it will be appreciated that the interconnect fabrication involves the following steps:

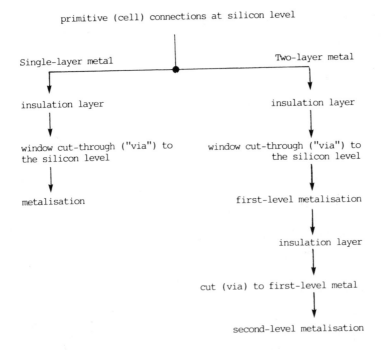

primitive (cell) connections at silicon level

Single-layer metal   Two-layer metal

insulation layer   insulation layer

window cut-through ("via") to   window cut-through ("via") to
the silicon level   the silicon level

metalisation   first-level metalisation

                                            insulation layer

                                            cut (via) to first-level metal

                                            second-level metalisation

Variants on this tabulation may be found, for example the two-layer polysilicon/single-layer metal P^{2}CMOS shown in Figure 2.39, but generally this represents most commercial interconnection structures.

It is clearly easier to route a masterslice array with two-layer metal than with single-layer, since routing constraints can generally be eliminated. The underlying polysilicon (see Fig. 2.34 for example), provides a further level of interconnect. The two metal layers are normally run at right-angles to each other, one layer running north-south on the chip floor plan, the other running east-west. The disadvantages with double-layer metalization, however, include:

1. Increased fabrication costs due to additional masks and fabrication steps
2. Yield problems due to metalization having to penetrate into deep contact cuts in order to make the required electrical connections

Single-layer metal, therefore, has economic attractions, but at the expense of potentially more difficult routing problems. Since single-layer metal still constitutes the principal technique with currently available masterslice arrays, we will consider this further. Additional details of two-layer metal may be found in published literature [51—56].

In general no single-layer metal interconnect should ever run over the top of the active silicon devices. A danger is that the voltage on a metalization track may act as a gate and form a parasitic field-effect transistor in the underlying silicon; if there is a very thick oxide insulation layer or multiple layers between the metal track and the active silicon level, then this is unlikely to be measurably significant,* but it still remains good engineering practice to route all metal where it can do not possible harm. Therefore for single-layer metal we will normally have a basic topology consisting of

1. Polysilicon "fingers" or short bars from the active devices or primitive cells into the wiring channels
2. Contact windows (vias) through the oxide insulation above the polysilicon, to provide contact points for the polysilicon-to-metal
3. Metal interconnect along the wiring channels, generally at right angles to the polysilicon fingers

This is illustrated in Figure 4.18a, the horizontal connections all being polysilicon, the vertical connections all being metal.

---

*A two-layer metal CMOS gate array has been reported, in which the second layer of metal interconnect is "folded over" the active cell area in order to reduce interconnection channel area [65]. This arrangement has been termed a "channel-less gate array."

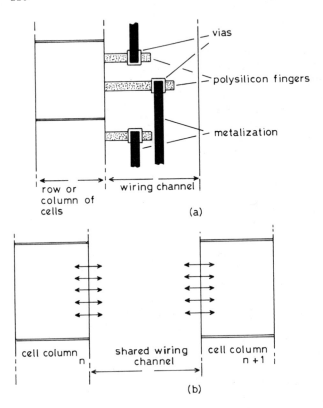

FIGURE 4.18   The polysilicon/single-layer-metal masterslice interconnection: (a) basic interconnection means; (b) shared wiring channels.

However, it is inevitable that connections will require to be made
from one column of cells to an adjacent column of cells, and therefore,
if a wiring channel is made exclusive to one column of cells all such
intercolumn connections will have to loop around the ends of the wiring
channels in the end-channel wiring space in order to be routed. To
avoid this unnecessary congestion in the end channels and long routing
lengths, it is common practice to have the intercolumn wiring channels
nonexclusive, that is, used by adjacent columns of cells, as indicated
in Figure 4.18b. This then raises a choice of options on how such
sharing shall be arranged.

Consider the routing situation shown in Figure 4.19a. Polysilicon
fingers may originate from cell positions which are exactly opposite
across the wiring channel, and reach only as far into the channel as
is necessary for them to connect to the required track. The routing
problem that then arises is how to assign connections to track positions

across the wiring channel such that polysilicon fingers at the same point along the wiring channel never overlap, and yet be able to connect from one cell column to the facing column, and to achieve this using the minimum number of track positions across the wiring channel. Hence there are constraints in two directions, namely neither the vertical metal interconnect nor the polysilicon fingers, respectively, must ever overlap.

There exists in the literature a number of algorithms which are designed to tackle this problem [57—61]. Nevertheless, it is possible that for a particular circuit and dedication of the cells (placement), these constraints may result in an impossible-to-route situation, such as shown in Figure 4.19b. Here the constraints are cyclical; the track for wiring net 1 must be to the right of that for net 2, track 2 must be to the right of track 3, but track 3 must also be to the right of track 1. The result is an impasse, in that net 3 cannot be routed with this placement [39]. Two possible solutions to this constraint are

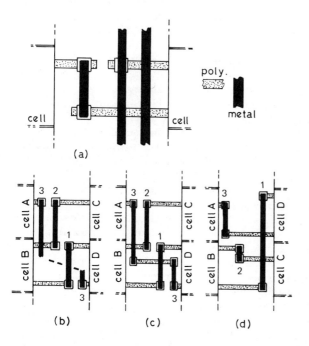

FIGURE 4.19 Polysilicon/metal routing in shared wiring channels: (a) connections between facing cells across the shared wiring channel; (b) an impossible-to-route constraint; (c) solution to (b), provision of additional wiring channel polysilicon fingers; (d) alternative solution to (b), re-placement of cells C and D.

shown in Figures 4.19c and d, although it must be appreciated that
these solutions may impose impossible-to-route constraints on yet further interconnections.

The use of additional polysilicon fingers as in Figure 4.19c to act
as underpasses in the wiring channels and thus allow dog-leg connections, may be found in many commercial gate-arrays. However, such
underpasses may not always be the full wiring channel width, and may
only be provided at occasional intervals, and hence are not necessarily
a perfect solution to the original constraint. A further problem with
the example illustrated in Figure 4.19 may also be observed, this being
that the lengths of polysilicon fingers from the cell connections are of
varied length; however since the polysilicon level is not normally a
variable part of a masterslice array these lengths cannot be altered to
suit the routing requirements, and thus the picture illustrated in Figure 4.19 is oversimplified (note that two-layer metal allows variations in
both east-west and north-south routing to be accommodated, but this
is not so with a fixed poly/single-layer metalization).

Hence, the problem of wiring crossings requires further consideration. Additionally, there is the problem of more easy access from
one wiring channel to an adjacent channel without having to loop
around the top or bottom via the end-channel wiring space.

Figure 4.20 shows several possibilities. Firstly, the cells in each
column may have both-sided access, as shown in Figure 4.20a. This
has several advantages, including: (1) the ability to swap connections
between cells in the same column from one wiring channel to the other,
thus easing congestion or constraints in a channel and equalizing track
density across the chip, and (2) the both-sided access allows through
connections to be made from one wiring channel to another. Alternatively, or additionally, polysilicon fingers which span the whole width
of the wiring channel(s) may be provided, as in Figure 4.20b. This
requires that the connections from cells on either side of a wiring channel must be staggered with respect to each other so as to be able to
interleave, but such a provision immediately eliminates the cyclic wiring
constraint of Figure 4.19b. Additional polysilicon fingers within the
wiring channel to allow dog-leg connections as in Figure 4.19c are not
now required, but the disadvantage is that the terminal pitch on the
standard cells may have to be increased in order to allow clearance
between the interleaved poly fingers. Figure 4.20c illustrates a combination of both architectural features.

We have shown single-cell columns in all these illustrations, and
have not specifically considered back-to-back cell columns as previously
illustrated in Figure 4.16c. Clearly both-side access is relevant only
for single-cell columns, but the arrangement of Figure 4.20b would be
relevant for back-to-back cell columns. Back-to-back architecture
may have advantages in reducing the total silicon area for the two rows
of cells due to possible sharing of power supply lines and other

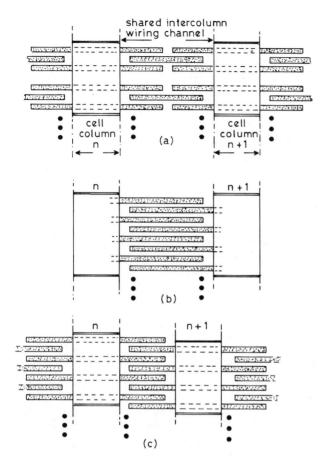

FIGURE 4.20   Further masterslice floor plan topology features:
(a) both-side access cells; (b) interleaved polysilicon fingers;
(c) both-side access plus interleaved fingers.

compaction, but precludes through routing from one wiring channel
to the next unless polysilicon underpasses are provided, or the cell
columns are occasionally split, such as shown in Figure 4.17f. On
balance there is no dominant case for back-to-back architecture.

The use of the polysilicon level for the interconnection dedication
of single-layer-metal gate arrays is therefore seen to be crucial. The
both-sided access of cell terminals illustrated in Figure 4.20a may also
be supplemented by separate polysilicon underpasses in each cell to
provide additional channel-to-channel routing. It is also possible to
"detach" the polysilicon fingers from the cell terminals, and connect

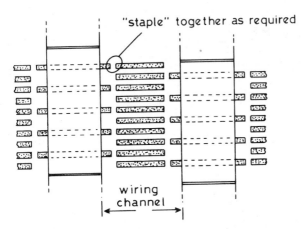

FIGURE 4.21   Variant on Figure 4.20c where each polysilicon finger
across the wiring channel has to be individually connected.

them only as required by metalization as shown in Figure 4.21, but
the additional vias and design detail required makes this a debatable
advantage—there is always the danger of omitting a "staple," and
circuit yield may be marginally affected in comparison with Figure
4.20.

The fundamental disadvantage with having to use polysilicon as a
connecting level, however, is that it has a resistivity many times that
of the aluminum metalization, and its use may degrade the maximum
speed performance of a circuit; a quoted figure is that driving through
six polysilicon underpasses is roughly equivalent to one additional gate
input loading on the driving circuit in CMOS technology [62], and
hence, it is good engineering practice to restrict the polysilicon level
use to the minimum, subject to other overriding considerations. Thus
the question of floor plan architecture and circuit routing to achieve
the highest possible on-chip speed is complex, since it not only involves the minimization of the polysilicon connections, but also the
lengths of the metal interconnect as well. This in turn becomes much
more a problem of placement of the circuit on the array than routing.

Research work is presently being undertaken to reduce the resistivity of polysilicon connections in order to improve circuit performance,
including the use of tantalum disilicide, titanium disilicide and other
compounds [63]. Also, as we have seen, the use of insulated substrate
materials in MOS technology considerably reduces the parasitic capacitances of wiring connections, and therefore, both of these trends will
help to minimize any adverse effects of polysilicon connections and
reduce the need for two-layer metal interconnect except where very
high performance is sought.

A final factor is also involved in masterslice design philosophy, this being whether fixed contact window sites (vias) to the polysilicon level shall be used, or variable sites for each custom dedication. In our preceding discussions with rows of cells separated by wiring channels and with polysilicon fingers in the wiring channels, it would be more usual to have variable via sites or multiple via sites per polysilicon finger, this being implicit in Figure 4.19b-d; the constraint in routing single-layer metal to fixed single via sites in such circumstances would be great. However, for masterslice component arrays such as illustrated in Figures 4.2, 4.7, and elsewhere, then fixed single via sites become more realistic, but at the expense of having to route each metal connection to the exact via position provided. Note that with fixed contact sites, the vias can be made as a part of the common masterslice fabrication, but with variable via sites an additional custom mask becomes necessary for chip dedication to define the via positions.

So far we have concentrated exclusively upon routing considerations in the intercolumn wiring channels. The peripheral end channels also require consideration, since constraints may arise where connections from intercolumn wiring channels and from chip I/O buffers require to cross. A common solution is to include in the end channels appropriate polysilicon fingers which may act as underpasses, the precise number provided and their positioning being somewhat arbitrary, as is the wiring capacity that shall be provided in these end channels. Note that underpasses may be required for all end-channel connections where they cross the dc power supply lines which feed the columns of cells, unless substrate-fed or other more complex forms of dc distribution to cells are provided.

If a routing procedure has to be followed which does not separate the end-column routing from the intercolumn routing, but requires the whole route from start to finish to be considered as one entity, then considerable complexity may accrue in automatic CAD or hand-routing procedures. However, if the chip topology is such that the intercolumn routing can be considered *separately* from the end-column routing, one not imposing any constraints on the other, then total routing is considerably simplified, because the two parts may be independently routed. What this entails is some form of termination philosophy at the ends of each intercolumn wiring channel, such that any track position from an intercolumn may connect to any track position in the end columns, which may involve some structure of polysilicon underpasses to allow crossovers. The number of wiring tracks which should be provided in end channels is very debatable; Rent's rule is no direct guide, as the number is strongly dependent upon I/O positioning, which may have to be fixed to suit a particular customer's pin-out requirements. A general rule-of-thumb would seem to be 1 to 2 times the wiring capacity provided in the intercolumn wiring channels.

All the above discussions are, in general, technology-independent, although possibly more relevant for masterslice uncommitted gate arrays rather than uncommitted component arrays. However, CMOS is a technology which has particular features that reflect in possible floor plan layouts. This has been discussed in Sec. 4.2.2, and generally illustrated in Figure 4.6, from which it will be recalled that the pairs of p- and n-channel devices present in CMOS logic circuits lend themselves to being laid out in rows (columns) of transistor-pairs. The most simple and flexible arrangement is equispaced single transistor-pairs, spaced such that all gates, sources, and drains (except those permanently connected to V_{DD} and V_{SS}) are available for interconnection purposes. This complete flexibility should be contrasted with other topologies as follows:

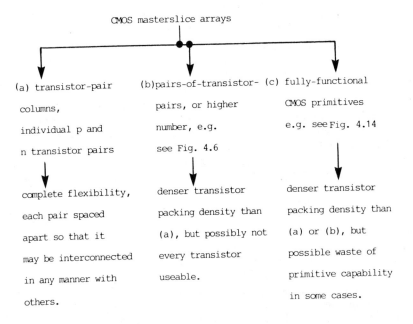

CMOS masterslice arrays

(a) transistor-pair columns, individual p and n transistor pairs

→ complete flexibility, each pair spaced apart so that it may be interconnected in any manner with others.

(b) pairs-of-transistor-pairs, or higher number, e.g. see Fig. 4.6

→ denser transistor packing density than (a), but possibly not every transistor useable.

(c) fully-functional CMOS primitives e.g. see Fig. 4.14

→ denser transistor packing density than (a) or (b), but possible waste of primitive capability in some cases.

Illustrations of transistor-pair masterslice arrays are given in Figures 4.22 and 4.23. Both of these commercial products will be seen to use separate polysilicon finger underpasses in the intercolumn wiring channels, with fixed multiple via sites per finger. The example in Figure 4.22 employs several finger lengths, with via sites on a rigid grid pattern; end-channel polysilicon fingers will also be noted. Metalization is constrained to run between the via grid spacing. The floor plan in Figure 4.23 again uses fixed multiple via sites per polysilicon finger, but all fingers are the full intercolumn wiring channel width.

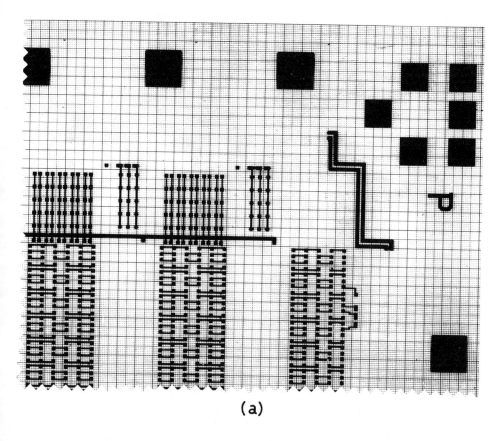

(a)

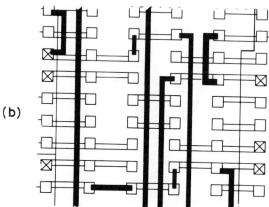

(b)

FIGURE 4.22 The floor plan interconnect layout of a CMOS master-slice transistor-pair array: (a) the polysilicon fingers in the inter-column wiring channels and in the end channel (transistor-pairs not shown); (b) enlarged detail of the intercolumn wiring channel, showing example metalization. (Details courtesy Micro Circuit Engineering, Ltd., Tewkesbury, England.)

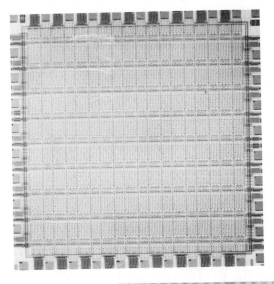

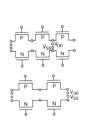

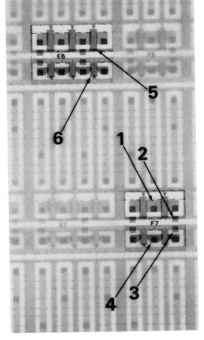

1. channel drain / source
2. V_{DD}
3. V_{SS}
4. n channel drain-
             source
5. p channel gate
6. n channel gate

FIGURE 4.23   The CMOS MCB Monochip floor plan of Interdesign. Illustrations are oriented to show rows rather than columns of *pn* transistor-pairs and wiring channels. (Details courtesy Interdesign, Inc., Sunnyvale, CA.)

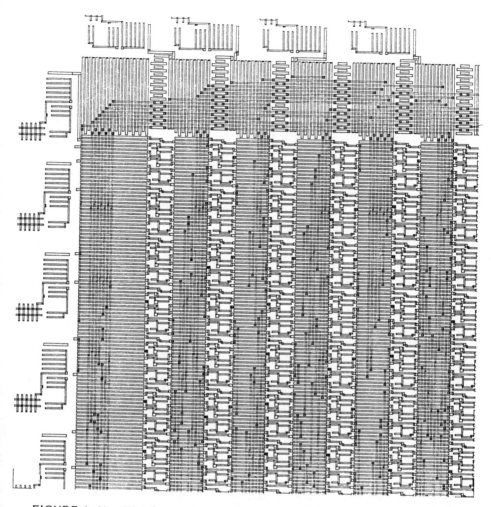

FIGURE 4.24   The floor plan architecture of an experimental CMOS
masterslice array with no cyclic routing constraints, using high-level
primitive cells and interleaved polysilicon fingers.   (Basic topology design University of Bath, detailed silicon layout Silicon Microsystems
Ltd., Malmesbury, England.)

   In contrast to these transistor-pair arrays, the floor plan layout of
an experimental CMOS masterslice array using the functional primitive
shown in Figure 4.14 is given in Figure 4.24.   This design uses both-
side access cells, as may have been discerned from Figure 4.14, with
variable position vias on the cell contact fingers.   No routing constraints

are present anywhere except for the available wiring channel track
capacities. The silicon area required for this layout, however, is fully
competitive with transistor-pair arrays, as may be found in available
literature [29—31].

Thus, to summarize there are very many, frequently conflicting,
considerations which have to be considered in the detailed routing
architecture of masterslice arrays. However, without doubt easily
routable topologies can be generated, which do not have any wiring
constraints except the maximum available wiring channel capacities.
The penalty for such ease of routing may be some increase in total chip
area and some reduction in absolute maximum on-chip performance, but
for the majority of commercial applications for masterslice arrays state-
of-the-art performance is not a significant customer requirement; a
performance equivalent of SSI/MSI discrete packaging is by far and
away the most usual customer requirement.

4.5 PLACEMENT CONSIDERATIONS

The question of placement of the required circuit on a masterslice array
topology prior to routing has not been addressed in the preceding discussions, although it must be undertaken before routing begins. Clearly the commitment of the individual cells of the array to specific duties
will affect the detailed routing of the chip, and may lead to a local density of interconnection higher than can be accommodated by the wiring
channel capacity. Conversely, with some initial placement choice, any
resulting bottleneck found during routing may hopefully be eliminated
by an appropriate rearrangement of the cell duties.

However, it will be appreciated that the statistical studies and Rent's
rule which we considered earlier do not require any a priori division of
the system into subdivisions and blocks per subdivision; the results
are statistically independent of the subdivisions, provided the number
of subdivisions is significant. Indeed, this is what is usually found
to happen in practice with the routing of uncommitted arrays. In general, provided the columns (rows) of cells are committed to their required duties in some sensible logical order, corresponding to the system circuit diagram, then no extensive sophistication is essential at
the preliminary design stage in an attempt to give some more advantageous placement on the chip. Clearly, if one were to allocate adjacent
cells to parts of the circuit diagram completely spaced away and electrically divorced from each other, one would expect to run into problems, but provided some sensible allocation is made, a reasonable initial
placement results. Some research has been done on the masterslice
array illustrated in Figure 4.24, in which an initial macro placement
allocation was perturbed and the resulting change in maximum channel

wiring density monitored for a given system using 114 of the 120 cells per chip; only minor changes in maximum wiring density were observed unless single cells were individually moved completely away from their "logical" position. Local peak wiring densities of course were modified by such placement perturbations, as would be expected. We will consider placement in more detail in Chapter 6, since it occupies a more important aspect in standard cell design than in masterslice array design, due to the desire to reduce wiring channel interconnection density to a minimum in the former to thus produce the smallest silicon area chip; with masterslice arrays, as long as the available wiring channel capacity is not exceeded there is normally no strong incentive to produce the "best" placement and routing design. As we will see later, there are a number of algorithms which have been published for automatic placement, involving for example, forces of attraction or repulsion between cells, where "force" is some logic connectivity or commonality, but no one algorithm is universally successful.

It is inevitable that with masterslice arrays, local routing limits must at times be exceeded with an initial placement. It is here that the supreme ability of the human operator comes into play, since he or she is far more efficient in recognizing possible re-placement solutions to such bottlenecks than any computer program. Human intervention therefore remains the best present means of solving such routing constraints.

4.6 CAD CONSIDERATIONS

The full design and fabrication process for a customized masterslice gate array is indicated in Table 4.5, where the interface between the original equipment manufacturer (customer) and the IC supplier (vendor) may occur at different levels of sophistication.

There are also considerably varied levels of CAD automation that may be present in this flow chart, ranging from very little or even none at all at the system design and layout levels, to increasingly sophisticated systems which aim to provide design verification, full timing analysis, and test-vector generation, as well as the fabrication requirements. Nevertheless, it is possibly true to say that less CAD may be present in the masterslice semicustom design area than in cell-library design, due partially to the nature of the masterslice approach with it predesigned but uncommitted chip, and to the fact that masterslice semicustom may be used for simpler and less sophisticated applications than cell libraries. We will consider the cell-library approach and its implications in detail in the following chapter.

Looking at currently available CAD tools for masterslice custom-specific design, a complex and to a large extent a disjointed picture is

TABLE 4.5 The Design and Fabrication Stages for a Masterslice Array, with Different Levels of Customer Involvement. (Based upon Rozeboom, R., Texas Instruments)

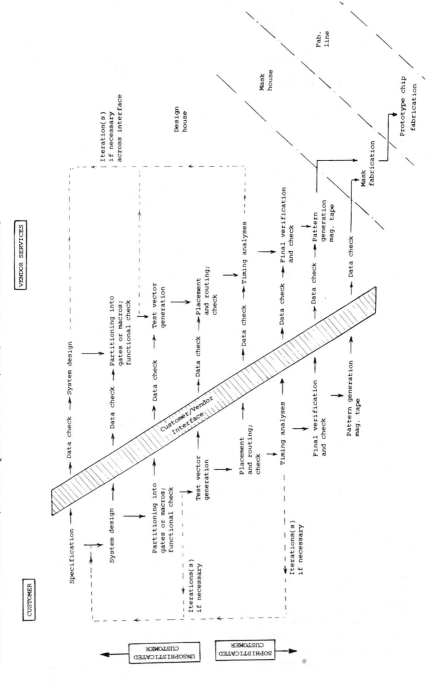

prevalent. The problem is compounded by the present lack of standardization of the formats in which data is handled by various possible software packages for placement and routing, logic verification, mask-pattern generation, and so on. This makes the assembly of a complete working CAD system by an original equipment manufacturer a difficult and often frustrating process, and means that each large IC manufacturer (vendor) tends to design and compile his own unique comprehensive design facilities.

The most clearcut CAD procedure for masterslice arrays is where all the design and fabrication details are under the control of the vendor, that is towards the top level of the picture given in Table 4.5. The vendor's CAD and support sophistication may then be very great, perhaps including international data links to serve customers globally remote from the mainframe computing facilities. Such a structure is illustrated in Figure 4.25, which caters for high-performance gate arrays with a choice of ECL, advanced low-power Schottky TTL, and CMOS technology. Customers may be given access to such a system in private for circuit entry and verification via a design entry terminal, but clearly there is a learning exercise then involved in teaching OEM designers about the system and its protocols.

Similar situations may be found elsewhere with other vendors. Figure 4.26 illustrates a further customer-vendor flow chart in which the customer may supply documentation only of his requirements, the documentation being: (1) a logic diagram of the required system and (2) a timing diagram for the system. The logic diagram supplied by the customer to the vendor msut be partitioned into logic cells or macros which correspond to the standard commitments available from the masterslice cells (3-input NANDS), and the total cell count must not exceed (approach?) the total number of basic cells available on the masterslice array.

Both of the above cases illustrate the isolation of the customer from the fine details and possible complexities of the CAD hardware and software. However, there may be a certain customer reluctance in plugging into a remote big-brother design system, and many would like to retain all design and verification stages under their own control, using the pattern-generation magnetic tape as the customer-vendor interface level (e.g., the bottom level of Table 4.5). Ideally, the customer (OEM designer) would like to undertake as far as possible his design activities in some vendor-independent form in order to give him multisource capability, the finalized design data being made vendor- and possibly technology-specific at the final interface level.

The recent introduction and rapid evolution of stand-alone CAD workstations is an important stride along this path; the terminology computer-aided engineering (CAE) is tending to be applied to this whole range of computer-based tools. However, it is currently true that the hardware capabilities of workstations have largely outstripped the available

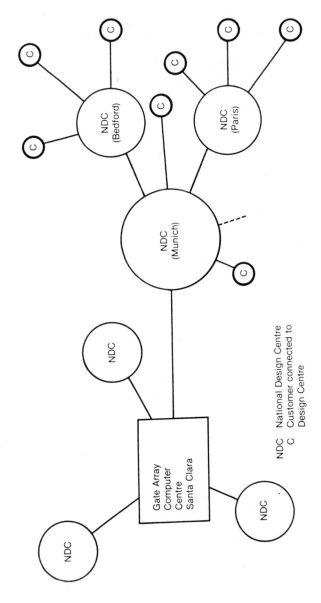

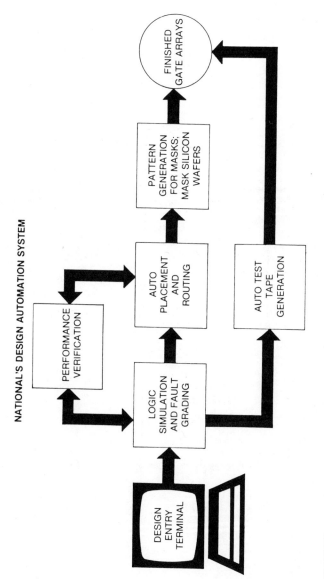

FIGURE 4.25 An internationally accessible CAD system for masterslice design. (Details courtesy National Semiconductor Corporation, Santa Clara, CA; further extensions to this network may now be available.)

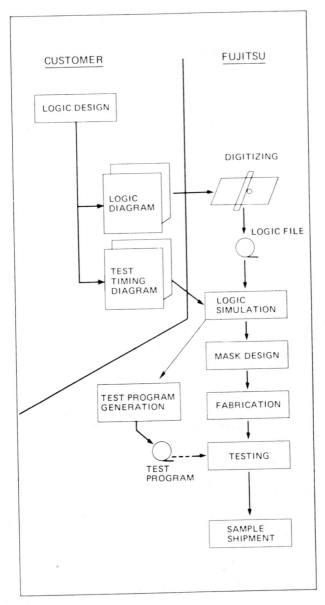

FIGURE 4.26   The customer-vendor interface of Fujitsu for masterslice arrays.   (Details courtesy Fujitsu Ltd., Kawaski, Japan.)

software packages required to form a fully integrated design system
in which design data from one phase of the design, (e.g., layout),
can pass freely to circuit simulation, placement and routing, timing
verification, and all other facets of the design procedure. The data
base management of stand-alone workstations and standardization of
data formats remains the outstanding need at the present time.

As we are at the moment specifically considering masterslice arrays,
and not the wider scenario of cell-library semicustom or full custom
design, there is no necessity in the CAD system for the manipulation
of shapes and geometries (polygon editing) such as is essential in full
custom design at the device level. Instead, the principal entry requirement into a CAD system is for drawing the circuit schematic,
which may run to several pages, and preparing net lists of the required
interconnections. The latter is the area where hand methods are very
prone to error, due to the ease by which the human designer can for-
get to list a connection or make some simple pin-numbering error. A
current CAE product, termed a "schematic designer," is shown in
Figure 4.27. This provides the OEM designer with a parts library
containing the most frequently required logic symbols, plus the facility
to create new ones as required, with the ability to start at a high level
of abstraction (e.g., large functional blocks with block-to-block interconnections), paging down to increasingly lower levels of design (e.g.,
macros, gates, or even transistor level if required). Hard-copy printing
of the complete design, together with the full net list and other inventory details are finally provided.

This then represents the modern equivalent of the drawing board
for system design purposes, and provides a relatively inexpensive
front-end entry to further CAD tools if required. However, this is
where machine interface problems may arise, due to the present wide
range of data formats and lack of overall standardization at the software level.

More comprehensive, and therefore expensive, stand-alone workstations for OEM use are becoming available. Figure 4.28 illustrates
one such product, which from its central data base can provide gate-
array placement and routing of specifically entered gate-array topologies. It may be noted that surrounding the data base are provisions
for

1. Front-end design capture, to provide interactive design entry,
   editing, and checking
2. Partitioning of the design and net listing
3. Functional simulation
4. MOS technology logic and timing simulation
5. Gate-array placement and routing, with one or two levels of
   custom interconnection
6. Standard cell placement and routing, again with one or two levels
   of interconnection

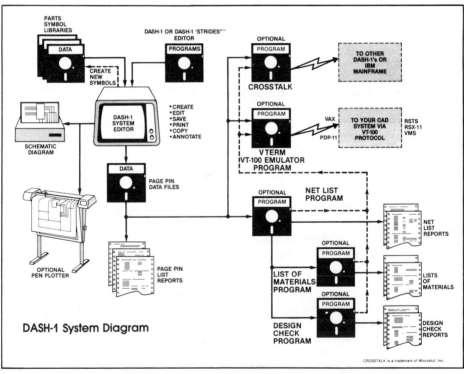

DASH-1 System Diagram

PARTS SYMBOL LIBRARIES

DATA

CREATE NEW SYMBOLS

DASH-1 OR DASH-1 'STRIDES'™ EDITOR

PROGRAMS

OPTIONAL

PROGRAM

CROSSTALK

TO OTHER DASH-1's OR IBM MAINFRAME

DASH-1 SYSTEM EDITOR

• CREATE
• EDIT
• SAVE
• PRINT
• COPY
• ANNOTATE

SCHEMATIC DIAGRAM

OPTIONAL

PROGRAM

VTERM IVT-100 EMULATOR PROGRAM

VAX

PDP-11

TO YOUR CAD SYSTEM VIA VT-100 PROTOCOL

RSTS RSX-11 VMS

DATA

PAGE PIN DATA FILES

OPTIONAL PEN PLOTTER

PAGE PIN LIST REPORTS

OPTIONAL

PROGRAM

NET LIST PROGRAM

NET LIST REPORTS

OPTIONAL

PROGRAM

LIST OF MATERIALS PROGRAM

LISTS OF MATERIALS

OPTIONAL

PROGRAM

DESIGN CHECK PROGRAM

DESIGN CHECK REPORTS

CROSSTALK is a trademark of Microstuf, Inc.

FIGURE 4.27   The electronic drawing-board for front-end entry of design data.   (Details courtesy FutureNet Corporation, Canoga Park, CA.)

228

Provision is made not only to use this workstation as a stand-alone facility, but also to interconnect a network of stations and other compatible computer systems in order to share resources and design details, and provide an integrated network design system with wide capabilities.

However, we are beginning to stray from the more simple areas of masterslice arrays into the wider area of CAD/CAE for cell-library and other custom-specific LSI design. Rather than duplicate material here, we will continue with further CAD/CAE details in the following chapter, including mention of the several software packages which are widely available for specific duties. Further details of workstations and CAD facilities may also be found in many available references [1,64,66—88].

(a)

FIGURE 4.28 Computer-aided engineering workstation to provide masterslice and other design facilities. (Details courtesy Silvar-Lisco, Palo Alto, CA.) (a) The workstation, (b) the available GARDS subsystem facilities, (c) interaction between the GARDS modules.

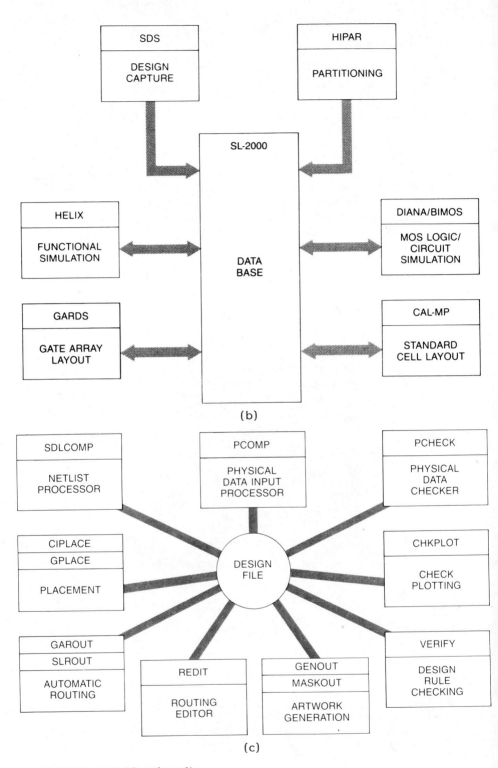

(b)

(c)

FIGURE 4.28 (Continued)

4.7 OEM COMMITMENT PROCEDURES

In spite of previous comments concerning the increasing availability
and resourcefulness of CAD tools, it is still common practice for
masterslice arrays to be manually placed and routed. This is particularly so where (1) one is considering uncommitted component arrays, such as illustrated in Figure 4.2, and (2) where the equipment
designer does not yet wish to invest in capital equipment for his
semicustom design due to relatively simple end products or uncertain
future requirements.

In the case of uncommitted component arrays, particularly those
with analog capability, an established technique is for the vendor to
supply the OEM design with large-scale layouts of the masterslice, for
example, ×100 or ×200, showing the positioning of all component items,
together with an appropriate design manual giving the necessary guidance, design rules, and performance specifications for the chip. Handrouting of the layout is undertaken, following which the master layout
is returned to the vendor for design rule checking, possible circuit
and timing verification, and fabrication. Figure 4.29 illustrates the
preparation of a typical UCA dedication interconnect mask from a handrouted master sheet. The nonorthogonal routing which is characteristic
of many component arrays will be noted, in contrast to the more structured orthogonal interconnections shown in following illustrations.

In the digital-only masterslice area, we find that by far the most
widely available type of masterslice which may be hand routed by the
customer is the single-layer metal CMOS technology transistor-pair
array, where the p- and n-channel pairs are laid down in rows (columns) as shown earlier (Figs. 4.6, 4.22, and 4.23). In such products,
the vendor will supply, say, ×100 master layout sheets, which show
the interconnection highways, available polysilicon underpasses, and
via sites, together with a range of transparant self-adhesive decals
which represent to scale the dedication of transistor-pairs to commonly
required digital building blocks. The latter macros may then be placed
on the master layout sheet as required by the system designer, and
hand routing between these macro I/Os is then undertaken. A very
close parallel between this and printed circuit board (PCB) design
methods may be noted.

Figure 4.30 shows the hand dedication of the transistor-pair masterslice previously illustrated (Fig. 4.22). The vendor again supplies
appropriate information on placement and routing, for example on
routing:

1. Metal tracks may cross any polysilicon underpass.
2. Only two metal tracks are allowed in the channel between contact vias.
3. Metal tracks are not allowed between adjacent contact vias.

FIGURE 4.29 Preparation of the ×200 interconnection mask for a bipolar uncommitted component array, made by overlaying a ×200 pencil
interconnection diagram prepared by the OEM designer. Only the
last of seven mask stages is custom-specific. (Details courtesy Interdesign, Inc., Sunnyvale, CA.)

4.   Metal tracks cannot cross each other (except by using a
     polysilicon underpass).
5.   Adjacent contact vias may be linked by a metal track.
6.   Metal interconnection tracks must not connect to unmarked
     contact windows on the standard cell decals (these points
     are connected to internal nodes).
7.   Metal tracks must not touch or go inside the boundaries of
     standard cells where these boundaries overlap the intercon-
     nection highway.

8. All tracks must be either vertical or horizontal.
9. Links between horizontal and vertical interconnection highway are made via polysilicon underpass.
10. The power rail runs in inner vertical channel of vertical interconnection highways (V_{SS} on right-hand side of chip, V_{DD} on left-hand side of chip).

The vendor would supply other detailed information as well. Such details are product-specific.

In addition to the manually placed-and-routed layout, the OEM designer must also supply the vendor with a logic diagram and test vectors for the design, plus a pin-out diagram showing bonding pattern requirements and a check sheet listing the inventory of macros and

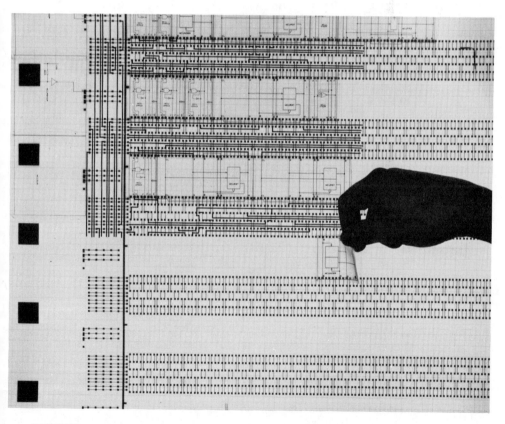

FIGURE 4.30 Manual placement and routing of the transistor-pair masterslice array illustrated in Figure 4.22. (Details courtesy Micro Circuit Engineering, Ltd., Tewkesbury, England.)

Table 4.6 The Vendor's Procedure on Reciept of a Logic Layout Such as Shown in Fig. 4.30

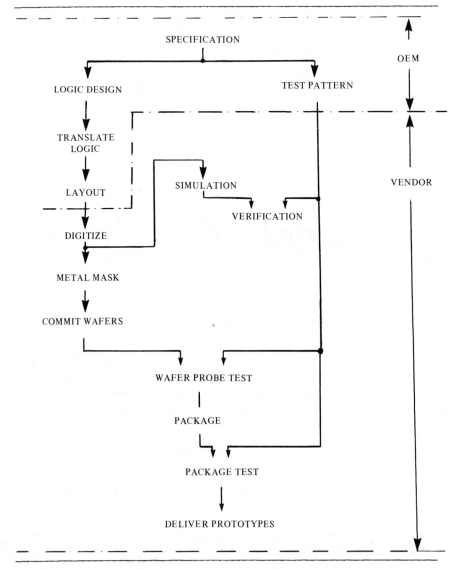

other housekeeping details. The vendor's procedure then follows the steps shown in Table 4.6.

Many transistor-pair masterslice vendors follow this design philosophy. Comprehensive design manuals, such as the ISO3/6 *Gate Array Design Manual* of Universal Semiconductor, Inc., may be available [89]. Such handrouting techniques may indeed be widely used by the vendors themselves for their own in-house activities, only becoming difficult when the number of transistor-pairs per die becomes very large, with thousands rather than hundreds of *pn* transistor-pairs.

Fully functional masterslice gate arrays may also be manually placed and routed in a similar manner, using ×100 or larger mastersheets. Again, transparent decals may be used to define the duties of cells or groups of cells (macros), leaving the principal design activity that of drawing in the required interconnect pattern. Routing of the master-slice array shown in Figure 4.24, which has no routing constraints, is shown in Figure 4.31.

The relatively few commercial masterslice products containing both analog and dedicated digital building blocks are usually available for customer routing, although the vendor may equally undertake the full design and fabrication if required. Some additional proposals to ease interconnection constraints between the on-chip analog and digital sections have recently been reported [90], and hence there may be further developments forthcoming in this rather specialized custom-specific area.

Nonmanual design procedures at OEM level are still evolving, and are difficult to summarize at the present time. However, without doubt we will see an increasing CAD/CAE involvement by the system designer, an area to which we will return in the following chapter.

4.8 COMMERCIAL STATE OF THE ART; COMMERCIAL AVAILABILITY

We have seen in Chapter 2 some general performance details of the various technologies which may be used in custom-specific microelectronics (in particular see Figs. 2.40, 2.41, and 2.42). In Chapter 6 we will look ahead to future trends and possibilities.

However, when dealing with customized circuits as distinct from volume-production standard parts, it is generally true to say that the very latest state of the art of the technology should not be used. Rather, a well-proven status should be involved, which will generate good yields to the customer without any fine tuning of the production line processes. Thus, the technically exciting publications given at specialized international conferences and symposia do not represent current availability for the average original equipment manufacturer, but may be forecasting viable situations three or more years ahead.

FIGURE 4.31 Manual routing of the masterslice array illustrated in
Figure 4.24. (Details courtesy University of Bath, Bath, England.)

As we have seen, masterslice arrays, either UCAs or UGAs, are currently available in all the major technologies, ECL, CML, I^2L, low-power Schottky TTL, STL, ISL, nMOS, and CMOS. Future evolution may see a disappearance of some of these bipolar variants, leaving possibly (1) ECL for the highest speed market, (2) STL or ISL for high current, low impedance applications, and (3) CMOS for the majority of applications, due to its good speed and packing density allied to low power dissipation. However, the OEM designer may at present have a choice of masterslice technology, the present status being generally shown in Figure 4.32. Nevertheless, he would be wise to choose a product which has been in production for at least a year if his equipment performance allows, with device geometries of not less than 5 microns, if at all possible. Device geometries of 3-2 µm are still not sufficiently well proven for the OEM designer to use with confidence of first-time success.

The number of industrial companies supplying custom-specific products is also rapidly evolving. It is characterized by an appreciable number of relatively small companies successfully competing alongside the major mass-market multinational IC manufacturers; indeed it would be true to say that the emergence of new companies has in many cases forced the multinationals to join the custom-specific market in order to secure their future existance. However, in such a growth situation

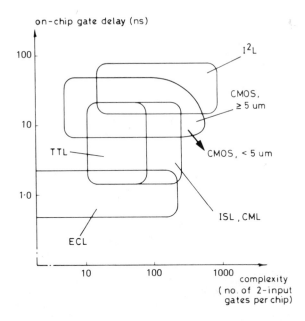

FIGURE 4.32   Masterslice technology capabilities presently available.

TABLE 4.7 Emergence of Industrial Companies Involved in Semi-custom Microelectronics from 1971 to 1983[a]

						Year
1971	1972	1973	1974	1975	1976	1977
Exar Ferranti Motorola	Harris	RCA	Integrated Microcircuits, Inc. Plessey	—	Philips	Interdesign Master Logic Microcircuits Technology, Inc. Signetics

[a]No guarantee of completeness or precise accuracy is made.

1978	1979	1980	1981	1982	1983
CDI	—	AMI	Barvon Research, Inc.	AMD	Alphatron
Heuer		Cherry	ESRO	AMCC	Citel
Motorola		CIC	Hitachi	Array Technology	Hughes
Racal		Eurosil	Holt, Inc.	Corintech	Monolithic Memories
		Integrated Circuit Systems	IMI	Custom MOS Arrays	NCM
		Micronas	LSI Logic	Fairchild	OKI
		Mital	MEDL	Fila Electronics	RIFA
		Nitron	Microcircuit Engineering, Ltd.	Fujitsu	STC-ATES
		Semi-processes, Inc.	National Semi	GIM	
			NCR	GTE	
			PMI	IMP	
			Rockwell Semiconductors Alcatel	Intersil	
			Sharp	Matra-Harris	
			Siemens	NEC	
			Seikosha	Raytheon	
			Telmos	Ricoh	
			Texas Instruments	Siliconix	
			Western Digital	SOREP	
			Zymos	STC Semiconductors	
				STL Microelectronics	
				Synertek	
				Thomson EPCIS	
				Toshiba	
				Universal Semiconductors	
				VLSI Technology	

a number of companies will not survive and already one encounters an
infant mortality rate among companies in this area, or discontinuation
of certain product lines.

Table 4.7 gives a summary of companies known to be involved in
semicustom microelectronics as of 1983. To this must be added a number of specialized design houses which act as professional middle-men
between a customer who requires a custom-specific circuit but is not
prepared or able to undertake the required design or specification details, and standard mask and fabrication line facilities.

Table 4.7 can be broken down into (1) masterslice component-array
suppliers, (2) masterslice gate-array suppliers, (3) other custom-
specific products such as cell-library. Any further breakdown into
fabrication technology, chip architecture, performance availability,
and other parameters would be an extremely complex endeavor. Comprehensive and very useful independent surveys may be found [1,3,
4,6,7,86,87,91,92], but in such a rapidly evolving area specific surveys may not be able to remain fully representative over a period of
time. We will not attempt any detailed company profile here, but refer
readers to surveys which continue to be published at frequent intervals. Table 4.8, however, gives an abbreviated summary of masterslice vendors, but this should be taken as representative only and not
at all globally comprehensive. Vendors not listed may have equally
viable products as those listed, with an increasing availability of CMOS
products from many vendors.

Certain generalized parameters may be extracted from the current
state of the art, such as relative manufacturing costs with differing
technologies, comparative silicon areas, percentage yield, and others.
Certain of these have been illustrated previously (Chap. 1, Figs. 1.7–
1.10); for additional information see detailed references [1,87,88,91,92].
However, one of the additional questions which will be of concern to
OEM designers is that of second source for their proposed custom-
specific product.

In general the more sophisticated the required product in terms of
logic per chip and maximum speed capability, the greater the difficulty
in second-sourcing. This is where the continuing evolution of OEM
workstations and software will eventually provide a breakthrough,
since a powerful aim and objective will be to evolve CAD/CAE resources
which are vendor-independent, the custom design being completed by
the OEM designer in some standardized data base form, from which
vendor-specific but fully interchangeable products may be produced.
However, until the independent development of such systems outside
the major semiconductor companies materializes, the OEM design will
be constrained to finding suppliers that have common semicustom products in order to provide second-source availability.

It is usual practice for second-sourcing to arise by licensing agreements between one vendor who has an available custom product and

TABLE 4.8   Masterslice Product Vendors[a]

Vendor	Bipolar arrays, with possibly analog capability (A)	MOS and CMOS arrays, no analog capability
Advanced Micro Devices	√	—
American Microsystems	—	√
California Devices	√ (A)	√
Cherry Semiconductors	√ (A)	—
Exor Integrated Systems	√ (A)	√
Fairchild	√	—
Ferranti	√ (A)	√
Fujitsu	√	√
Harris Semiconductors	√	√
Holt Integrated Circuits	√ (A)	√
Hughes	—	√
Integrated Microcircuits	√ (A)	√
Interdesign	√ (A)	√
LSI Logic	√	√
Master Logic	—	√
Micro Circuit Engineering (UK)	—	√
Motorola	√	√
National Semiconductor	√	√
Nitron	—	√
Plessey	√ (A)	√
Raytheon	—	√
RCA	—	√
Signetics (Mullard)	√	√
Texas Instruments	√	√
Western Digital	—	√

[a]This listing is not comprehensive or guaranteed to be currently accurate (see text).

TABLE 4.9   Some Present Known Second-Sourcing Agreements[a]

Original source	Licenses	Product
Applied Micro Circuits	Thomson CFS, Signetics	ECL
California Devices	AMI, LSI Logic, Telmos	sg.CMOS
Exar	Cherry, Silicon Arrays	TTL Analog
Ferranti	Interdesign	CML, sg.CMOS
Fujitsu	Silicon Arrays	sg.CMOS
Interdesign	Ferranti, Exar	TTL Analog
LSI Logic	AMD, National	ECL
Master Logic	Eurosil, Exar, HMT, Interdesign, Monosil, Nitron, Telmos	mg.CMOS
Motorola	LSI Logic, National	ECL
Micro Circuit Engineering	MEDL, GIM (UK), Plessey, ITT.	sg.CMOS
Toshiba	LSI Logic	sg.CMOS
Universal Semiconductor	Nitron	sg.CMOS
Harris	Texas Instruments	sg.CMOS
Texas Instruments	Harris	Schottky TTL

[a]Not guaranteed to be complete or currently accurate.

another vendor with appropriate experience, but does not have a directly competing product. It is not usual for a new product item to be jointly funded and developed. Table 4.9 gives some presently known second-sourcing agreements for masterslice die [1,87,88,91,93], but again the global situation is continuously changing. Additionally, the customer should also consider how far a "second source" is truly independent from the original source of the product; does the supplier for example (1) fabricate the masterslice wafers complete, (2) stockpile supplies from the original source and undertake the custom dedication on the wafers, or (3) merely act as an intermediate design house, referring all fabrication dedication back to the original supplier? Indeed practical problems can arise when separate organizations attempt to undertake the metalization commitment of masterslice arrays not of their own manufacture, since the accuracy of step-and-repeat photolithography is such that "run-out" across the wafer can occur, that is, the exact geometric spacings and dimensions are not precisely uniform across the whole wafer due to lens distortion and other imperfections. Hence, should metalization masks be made by different camera and mask-making equipment, exact match between uncommitted wafer and metalization will not occur [94]. One method of ensuring correct wafer match is of course for all suppliers to use the same camera and mask-making

TABLE 4.10 The Decision/Questionnaire Tree Which Must Be Addressed by the OEM Designer for Custom-Specific Circuits

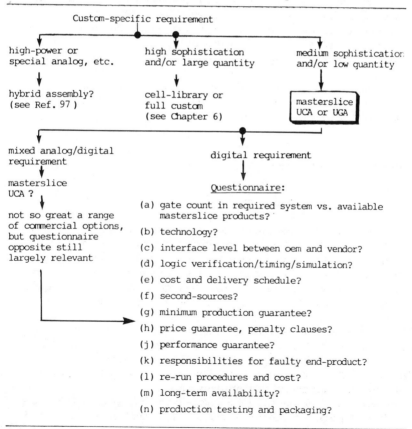

Custom-specific requirement

high-power or special analog, etc. → hybrid assembly? (see Ref. 97)

high sophistication and/or large quantity → cell-library or full custom (see Chapter 6)

medium sophistication and/or low quantity → masterslice UCA or UGA

mixed analog/digital requirement → masterslice UCA ? → not so great a range of commercial options, but questionnaire opposite still largely relevant

digital requirement

Questionnaire:

(a) gate count in required system vs. available masterslice products?

(b) technology?

(c) interface level between oem and vendor?

(d) logic verification/timing/simulation?

(e) cost and delivery schedule?

(f) second-sources?

(g) minimum production guarantee?

(h) price guarantee, penalty clauses?

(j) performance guarantee?

(k) responsibilities for faulty end-product?

(l) re-run procedures and cost?

(m) long-term availability?

(n) production testing and packaging?

equipment, so that imperfections remain constant, but this rather defeats the issue of second-source availability.

Hence, we see that there are many practical decisions facing the OEM designer in the use of masterslice arrays, not the least being technology obsolescence and parts obsolescence. As the custom-specific microelectronics industry matures, these problems should decrease, with the emphasis becoming more upon standardization and portability of the software design aids for commitment design procedures. Moves toward the standardization of MOS and CMOS design rules may also be observed [95,96], which have a particular merit of attempting to establish common ground between different IC manufacturers rather than push the state of the art to beyond what is readily achievable.

The commercial state of the art and availability of workstations and other CAD/CAE hardware and software will be deferred to the following chapter. Hence, we will conclude this discussion by attempting to summarize in Table 4.10 a typical decision tree. Note that we have not had occasion to discuss special applications which may arise in which it is relevant to consider hybrid assemblies. Such assemblies may be particularly relevant for specialized applications requiring mixed analog and digital capability allied with high power requirements, as may be encountered in electromechanical control systems. Details of hybrid microelectronics may be found elsewhere [97], being outside the mainstream of our present interests.

4.9 PACKAGING AND TESTING

Packaging of custom-specific circuits follows the same methods as off-the-shelf volume production products. There is no obvious necessity to develop packaging specifically for the custom market if standard products are available and appropriate, although many small-volume multichip substrate carrier products are available, usually ceramic substrate with dual-in-line package pins, which are appropriate for the mounting of several die per final package assembly [97]. Specialist companies, such as Augat, Inc., and others cater for many varied requirements.

However, the broad spread of "standard" packaging for integrated circuits is given in Table 4.11. Figure 4.33 illustrates some of these. Figure 4.5 also included an illustration of an 84-pin flat package used in Schottky-TTL custom products. By far the most commonly employed product is the familiar but not entirely satisfactory dual-in-line package (DIL or DIP), with 0.100 in. (2.54 mm) spacing of the package pins, and widths requiring typically 0.300 in. (7.62 mm) or 0.600 in. (15.24 mm) pitch of the rows of mounting holes. This is illustrated in Figure

TABLE 4.11 The General Hierarchy of Packaging for Custom and Semicustom ICS, Omitting Special Structures for Particular Applications

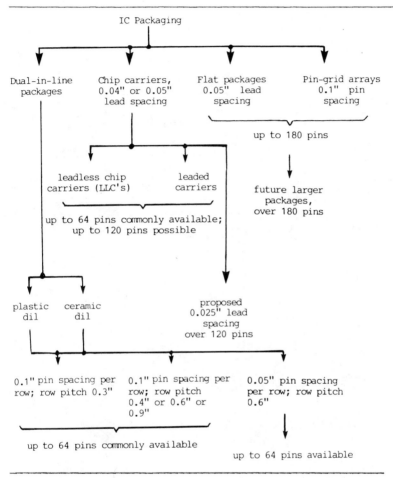

4.34. Such packaging is reasonably suitable for up to approximately 40 pins per package. Beyond this size insertion and removal problems as well as the floor area of the package becomes an embarrassment. Also, as pin count increases, the ratio of the longest chip-to-pin bonding wire to shortest bonding wire escalates, being about 6:1 for a 40-pin package; this causes not only mechanical problems but also different lead resistance and inductance values, leading to timing problems with very high speed circuits.

The principal trends and difficulties which are present in IC packaging are as follows:

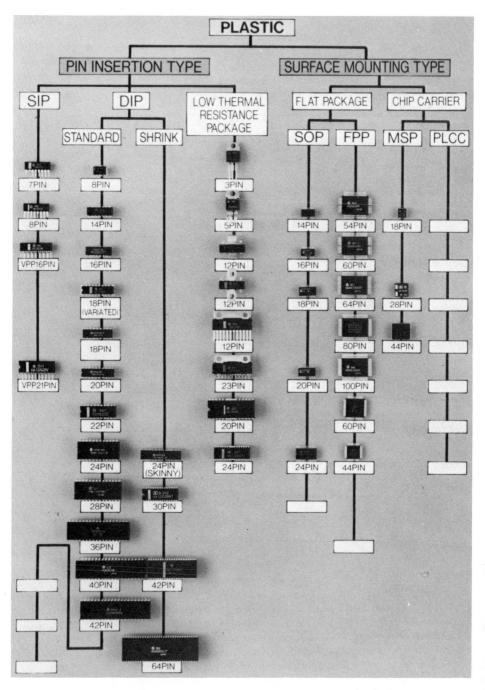

FIGURE 4.33 Typical present-day IC packages. (Details courtesy Hitachi Electronic Components, UK.)

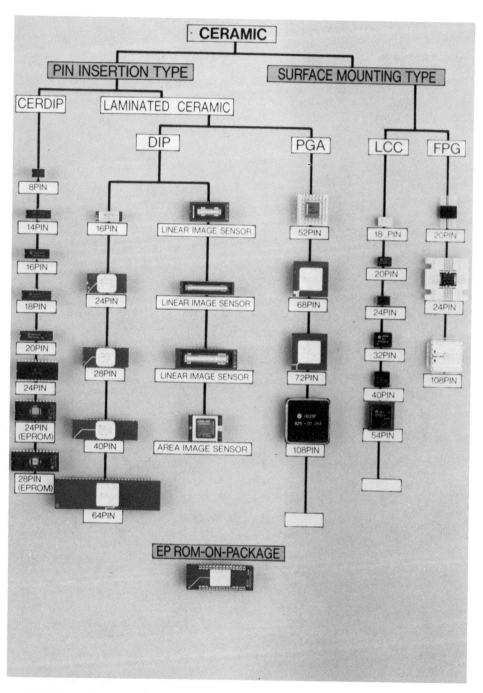

FIGURE 4.33 (Continued)

plastic packages

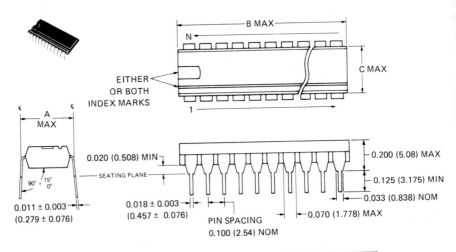

PINS DIM.	16	18	20	22	24	28
A (MAX)	0.325 (8.255)	0.325 (8.255)	0.325 (8.255)	0.425 (10.80)	0.625 (15.88)	0.625 (15.88)
B (MAX)	0.870 (22.1)	0.920 (23.37)	1.070 (27.18)	1.120 (28.45)	1.270 (32.26)	1.440 (36.58)
C (MAX)	0.270 (6.858)	0.270 (6.858)	0.270 (6.858)	0.355 (9.017)	0.550 (13.97)	0.550 (13.97)

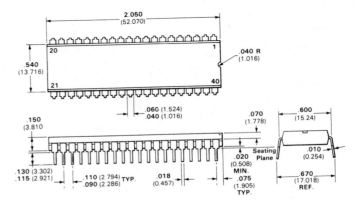

FIGURE 4.34   The dimensional details of standard dual-in-line packages.

ceramic packages

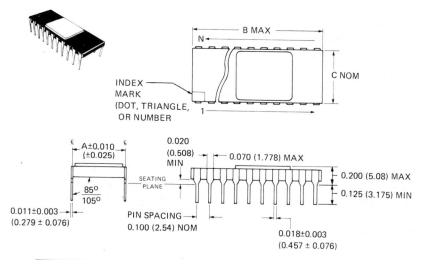

PINS DIM.	16	18	20	22	24	28
A ± 0.010 (± 0.025)	0.300 (7.62)	0.300 (7.62)	0.300 (7.62)	0.400 (10.16)	0.600 (15.24)	0.600 (15.24)
B(MAX)	0.840 (21.34)	0.910 (23.11)	1.020 (25.91)	1.100 (27.94)	1.290 (32.77)	1.415 (35.94)
C(NOM)	0.295 (7.493)	0.295 (7.493)	0.295 (7.493)	0.395 (10.03)	0.595 (15.11)	0.595 (15.11)

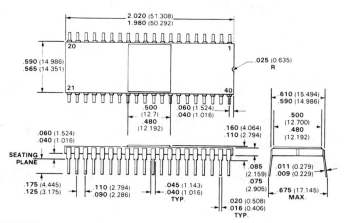

FIGURE 4.34   (Continued)

1. An increasing number of pins required per package
2. An increasing heat dissipation problem, not only within the packages, but also on the mounting boards which have to assist in dissipating the heat generated—this is particularly serious with large high-speed ECL chips
3. An increasing difficulty in easy mounting and removal of packages on/from mounting boards
4. An increasing need for smaller packages in order to reduce the length of bonding between chip and package pins, and hence improve signal propagation times

All these problems tend to favor the development and use of flatpacks, chip carriers, and pin-grid arrays over dual-in-line packages. In particular, the leadless chip carrier (LLC), in which the chip cavity is a high percentage of the overall carrier area and where the connections are on all four sides of the carrier, represents possibly the most useful alternative to the DIL for general purpose OEM use. Its leads are short and bent under the packages edges which allows a complete package to be soldered to the surface of a mounting board without the difficulties of insertion of pins through the board. Nevertheless, problems still remain, such as differential expansion between the LCC and board during assembly, and lead solderability [107]. The more expensive flatpacks and pin-grid-array carriers may not suffer so severely from thermal mounting stresses, but in general tend to be much more expensive [104].

Further details of packaging problems and developments may be found in published literature [98—106]. With the very high reliability and performance now present at the working chip level, the packaging and interconnection level is increasingly becoming the controlling factor in overall system reliability and performance. Hence, it is in this area that many worthwhile developments will continue to unfold.

Finally, testing. This is often the Cinderella and last detail considered by the OEM designer, and is frequently the cause of problems between customer and vendor of custom-specific products. With simple masterslice arrays the problems may not be severe, and a comprehensive test specification can be provided and undertaken. However, with increasingly large custom-specific circuits, such as those that requiring the more complex and higher pin-count packages such as we have just been considering, considerable complexity in defining and executing appropriate acceptance testing may be present [108].

This problem is equally if not more serious in the cell-library custom-specific area, and we will defer further discussion of this subject until the following chapter.

4.10 SUMMARY

We have attempted in this chapter to review both the principles and current practice relating to the wide range of masterslice semicustom products. It has been shown that a very wide range of products are available for OEM use, with every major IC technology well represented. At the simplest conceptual level, the masterslice approach represents the assembly and interconnection on one silicon die of a limited range of standard parts required to form a system, directly comparable to the assembly on a printed circuit board of discrete components or functional primitives with which all original equipment manufacturers will be familiar.

As we have seen, the high technology which is inevitably involved in fabrication and also in optimum design is a source of many present problems. Such problems are inevitable in a growth situation where continuing developments are taking place at the silicon level and at the system and circuit design level. The problems of most concern at present are:

1. Documentation from vendor to customer to describe clearly what is the masterslice product, its topology, and its full commitment procedure.
2. Standards between vendors to define clearly to the customer what is the capacity and capability of the product. Statements such as 500 gate-equivalents, and gate delays of 10 ns are not meaningful with detailed clarification.
3. Second-sourcing and precisely what is provided by a second source, whether full fabrication or otherwise.
4. Packaging and testing, particularly how this affects the larger size products.
5. Finally, and possibly the one of least clarity at the moment, how CAD and CAE developments will affect the subject and the customer/vendor interface.

In spite of all these problematic areas, masterslice products will continue to be used in increasing volume, and will not be superceded by cell-library or other custom-specific techniques.

REFERENCES

1. Source III Inc. *Gate Arrays: Implementing LSI Technology.* Electronic Trend Publications, Cupertino, CA, 1982.
2. Groves, W. Do it yourself VLSI. *Digital Design*, 7, September 1982.

3. Sziron, S. Picking the right custom i.c. solution, *Defense Electronics*, 96-101, September 1982.
4. Hartmann, R. F. Design and market potential for gate arrays. *Lamba*, 1(3):56-59, 1980.
5. Jack, M. A., Semicustom i.c. design in the UK—a capability profile. *IEE Electronics and Power*, 29:217-221, 1983.
6. Product Focus. Custom electronics. *Electronic Engineering*, 54(667):50-71, 1982.
7. Product Focus. Logic arrays—what's on offer. *Electronic Engineering*, 54(668):57-60, 1982.
8. Supplement. Custom design techniques. *Electronic Engineering*, 55(678):99-138, 1983.
9. Rappaport, A. EDN Semi-custom i.c. directory. *EDN*, 28(4): 78-218, 1983.
10. Twaddell, W. Uncommitted i.c. logic. *EDN*, 25(7):88-98, 1980.
11. Huffman, G. D. Gate-array logic. *EDN*, 26(19):86-94, 1981.
12. Hicks, P. J. *An Overview of Semi-Custom I.C. Technologies.* Pre-Conference Tutorial, Proc. 3rd. Int. Conf. on Semi-custom I.Cs, London, October 1983.
13. Grierson, J. R. Modern developments in gate arrays. *IEE Electronics and Power*, 28:244-248, 1982.
14. Braeckelmann, W. Custom made integrated circuits. *Proc. Int. Conf. on New Trends in Integrated Circuits*, Paris, 99-107, April 1981.
15. Hollock, S. Semi-custom IC—VLSI made easy. *Proc. Int. Conf. on New Trends in Integrated Circuits*, Paris, April 1981.
16. Posa, J. G. Gate arrays—a special report. *Electronics*, 53(21): 145-158, 1980.
17. Tobias, J. R. LSI/VLSJ building blocks. *IEEE Computer*, 14(8): 83-101, 1981.
18. Yano, S., Ouchi, Y., Kimura, K., and Ito, Y. A masterslice approach to low-energy high-speed lsi. *Proc. IEEE Int. Conf. on Circuits and Computers*, 203-206, 1980.
19. Chin, W., Chen, J., and Jen, T. S. A high density cost performance bipolar masterslice for low end systems. *Proc. IEEE Int. Conf. on Circuits and Computers*, 197-202, 1980.
20. Cox, A. M. A bipolar lsi gate array using current-mode logic. *Proc. IEEE Int. Conf. on Circuits and Computers*, 215-218, 1980.
21. Special Issue. Semi-custom ICs. *Microelectronics J.*, 14(3): 1983.
22. Lipp, R. Understanding gate arrays ensures wise chip selection. *EDN*, 26(19):99-103, 1981.
23. Dantee, A., and Desuche, J. High speed CMOS gate arrays. *Microelectronis J.*, 14(3):41-52, 1983.
24. Edwards, C. R. A special class of universal-logic-gates and their evaluation under the Walsh transform. *Int. J. Electronics*, 44:49-59, 1978.

25. Chen, X., and Hurst, S. L. A comparison of universal-logic-modules and their application in the synthesis of combinational and sequential logic networks. *IEEE Trans. Computers*, C.31: 140-147, 1982.

26. Muzio, J. C., Miller, D. M., and Hurst, S. L. *Spectral Techniques in Digital Logic*. Academic Press, London, 1985.

27. New, A. M. Statistical efficiency of universal logic elements in the realisation of logic functions. *Proc. IEE*, 129E: 93-99, 1982.

28. Chen, X., and Hurst, S. L. A consideration of the minimum number of input terminals on universal logic gates and their realisation. *Int. J. Electronics* 50(1):1-13, 1981.

29. Hurst, S. L., and Jennings, P. I. Layout and design criteria for routable masterslice gate arrays. *Proc. IEEE Custom IC Conf.*, 322-326, 1983.

30. Jennings, P. I., McDonald, A., and Hurst, S. L. An exercise into a highly routable masterslice gate array topology. *J. Semi-Custom ICs*, 1(3):5-18, 1984.

31. Jennings, P. I. A topology for semicustom array-structured lsi devices and their automatic customisation. *Proc. IEEE 20th Design Automation Conf.*, 675-681, 1983.

32. Exar. *Custom and Semicustom Product Guide*. Exar Integrated Circuits Inc., Sunnyvale, CA, 1981.

33. Lau, S. Y. Integrated Schottky logic gate array. *Electronic Components and Applications*, 2(2):1980. (Reprinted as Mullard/Signetics Technical Note 143, 1980.)

34. Refiogla, H. I. A CMOS gate array family concept makes CMOS systems a reality. *Proc. 3rd Int. Conf. on Semi-Custom ICs*, paper 2/13, London, 1983.

35. Landman, B. S., and Russo, R. L. On a pin versus block relationship for partition of logic graphs. *Trans. IEEE*, C.20: 1469-1479, 1971.

36. Heller, W. R., Mikhail, W. F., and Donath, W. E. Prediction of wiring space requirements for lsi. *IEEE Proc. 14th Design Automation Conf.*, 32-42, 1977.

37. Jennings, P. I. A discussion document on masterslice layout. University of Bath, School of Electrical Engineering, Wolfson Report No. 11, March 1982.

38. Jennings, P. I., Hurst, S. L., and McDonald, A. A highly-routable ULM gate-array and its customisation. *Trans. IEEE*, C.33:22-34, 1984.

39. El Gamel, A. Stochastic models for interconnections in integrated circuits. *Proc. IEEE Int. Conf. on Circuits and Computers*, 77-79, 1980.

40. McDonald, A., and Jennings, P. I. A microelectronic gate array designed to give efficient automated circuit performance. *IEE Proc. Electronic Design Automation Conf*, 62-70, 1984.

41. Jennings, P. I. A comparison of the area requirements of a variety of gate array topologies together with some tentative conclusions. University of Bath, School of Electrical Engineering, Wolfson Report No. 9, November 1981.
42. Weinberger, A. Large scale integration of MOS complex logic: a layout method. *IEEE J. Solid-State Circuits*, SC.2:182–190, 1967.
43. Lallier, K. W., Hickson, J. B., and Jackson, R. K. A system for automatic layout of gate array chips. *Proc. European Conf. Electronic Design Automation*, 54–58, 1981.
44. Crispin, R. J. Computer-aided-design for the STL-83 ULA. *Proc. European Conf. Electronic Design Automation*, 59–62, 1981.
45. Mole, G. F., Alshahib, I. A., and Kinniment, D. J. A CAD verification system for ULA designs. *Proc. European Conf. Design Automation*, 49–53, 1981.
46. Burkard, W. D. Semi-custom lsi at Storage Technology Corporation. *VLSI Design*, 2(3):14–18, 1981.
47. Gould, J. M., and Edge, E. M. The standard transistor array STAR: a two-layer metal semicustom design system. *Proc. IEEE 17th Design Automation Conf.*, 108–113, 1980.
48. Shiraishi, H., and Hirose, F. Efficient placement and routing techniques for masterslice lsi. *Proc. IEEE 17th Design Automation Conf.*, 458–464, 1980.
49. Vernon, J. A., Crocker, N. R., and Singleton, J. P. A ULA is more than silicon. *Proc. European Solid-State Circuits Conf.*, 51–55, 1979.
50. Katz, D. Gate array chips and computer aided design methods for custom lsi/vlsi. *Proc. IEEE Int. Conf. on Circuits and Computers*, 207–210, 1980.
51. Hollock, S. The advantages and cost/yield trade-offs of multiple layers of metalisation. *Proc. 2nd Int. Conf. on Semi-Custom ICs*, paper 3/6, London, 1982.
52. Van Nielen, H. Safeguarding speed performance in CMOS gate arrays. *Proc. 3rd Int. Conf. on Semi-Custom ICs*, paper 3/8, London, 1983.
53. Adamson, R. W. A new family of high-speed two-layer interconnect arrays designed for automated placement and routing. *Proc. 3rd Int. Conf. on Semi-Custom ICs*, paper 3/9, London, 1983.
54. Dantec, A. AGATE MHS high-speed gate arrays supported by efficient design automation. *Proc. 3rd Int. Conf. on Semi-Custom ICs*, paper 3.10, London, 1983.
55. Ting, B. S., and Tien, B. N. Routing techniques for gate array. *IEEE Trans. on Computer Aided Design*, CAD2:301–312, 1983.

56. Tsukiyama, S., Harada, I., Fukie, M., and Shirakawa, I. A new global router for gate array lsi. *IEEE Trans. on Computer Aided Design*, CAD2:301–312, 1983.

57. Yoshimura, T., and Kuh, E. S. Efficient algorithms for channel routing. *IEEE Trans. Computer Aided Design of Circuits and Systems*, CAD1:25–35, 1982.

58. Kernighan, B. W., Schweikert, D. G., and Persky, G. An optimum channel routing algorithm for polycell layouts of integrated circuits. *Proc. IEEE Design Automation Workshop*, 50–59, 1973.

59. Hashimoto, A., and Stevens, J. Wire routing by optimising channel assignment within large apertures. *Proc. IEEE Design Automation Workshop*, 155–169, 1971.

60. Rivest, R. L., and Fiduccia, C. M. A "greedy" channel router. *Proc. IEEE 19th Design Automation Conf.*, 418–424, 1982.

61. Burstein, M., and Pelavin, R. Hierarchical Channel Router. *Proc. IEEE 19th Design Automation Conf.*, 591–597, 1983. *Integration, The VLSI Journal*, 1(1):21–38, 1983.

62. American Microsystems: Designing with gate arrays, Part 3. American Microsystems Inc., CA. A Training Course, 3 parts, 1983.

63. Davis, R. D. The case for CMOS. *IEEE Spectrum*, 20:26–32, October 1983.

64. Rozeboom, R. Capable software tools ease semicustom IC design. *EDN*, 28(10):185–193, 1983.

65. Readout Feature. Channel-less gate array reduces die size. *Electronic Engineering*, 55(678):113, 1983.

66. Rappaport, A. Computer-aided-engineering tools. *EDN*, 28(19):106–138, 1983.

67. Rappaport, A. Evolving CAE workstations furnish increased sophistication. *EDN*, 28(8):33–44, 1983.

68. Koford, J. S., and Jones, E. R. A development system for gate arrays. LSI Logic Corporation, Milpitas, CA, Reprint No. L23, 1981.

69. Garcia, S., and Sriram, K. S. A survey of IC CAD tools for design, layout and testing. *VLSI Design*, 3(5):68–73, 1982.

70. Hardage, C. Trends in gate array technology. *Microelectronics J.*, 14(4):5–10, 1983.

71. Various papers. *J. Semi-Custom ICs, Special Issue on Workstations*, 1(4):1984.

72. Claiborne, J. Automated workstation fills CAD gap. *Electronic Design*, 30(2):193–201, 1982.

73. Jenné, D. C., and Stamm, D. A. Managing VLSI complexity: a user's viewpoint. *VLSI Design*, 3(2):14–20, 1982.

74. National Semiconductor. Gate arrays and design automation. Publicity brochure, National Semiconductor Corporation, Santa Clara, CA.

75. Miller, D., and Miranker, G. Terminal-based engineering system cuts logic time tenfold. *Electronics*, 55(18):135–139, 1982.

76. Miller, D., and Rubin, J. Structured logic design system is fast and affordable. *Electronics*, 54(23):117–120, 1981.

77. Various papers. Proc. 2nd and 3rd International Conf. on Semi-Custom IC's, London, November 1982 and November 1983.

78. Texas Instruments. Texas Instruments Logic Array Design System (TILADS). Publicity brochure, Texas Instruments, Inc., Houston, Texas.

79. Comerford, R. High-level graphics tools master data interpretation. *EDN*, 28(26):40–46, 1983.

80. Robson, G. Benchmarking the workstations. *VLSI Design*, 4(2):58–61, 1983.

81. Broster, D., and South, A. The role of CAD in semicustom design. *IEE Electronics and Power*, 29:42–46, 1983.

82. Angus, C. J. Interactive graphics systems for CAE. *IEE Electronics and Power*, 29:13–16, 1983.

83. Hellier, W. E. The computer-aided engineering workstation. *IEE Electronics and Power*, 29:69–71, 1983.

84. Harrow, P. The selection of CAD systems. *IEE Electronics and Power*, 29:63–68, 1983.

85. Werner, J. Sorting out the workstations. *VLSI Design*, 4(2):46–55, 1983.

86. Trowbridge, M. Integrated design automation. *Electronic Product Design*, 4:51–54, May 1983.

87. ICE Report. *Status '81: A Report on the Integrated Circuit Industry*. Integrated Circuit Engineering Corporation, Scottsdale, AZ, 1981.

88. ICE Report: *Status 1984: A Report on the Integrated Circuit Industry*. Integrated Circuit Engineering Corporation, Scottsdale, AZ, 1981.

89. Universal Semiconductor. *The ISO3/6 Gate Array Design Manual*. Universal Semiconductor Inc., San Jose, CA, 1982.

90. Kemp, A. J. Architectural considerations for an uncommitted circuit intended for combined digital and analogue functions. *Microelectronics Journal*, 14(3):21–30, 1983.

91. Report. *The Impact of Gate Array Technology on IC Components*, Strategic, Inc., San Jose, CA, 1983.

92. Report. *The Impact of Custom Circuit Alternatives on IC Components*. Strategic, Inc., San Jose, CA, 1982.

93. News Item. Gate array manufacturing cost/yield. *VLSI Design*, 3(5):10, 1982.

94. Balch, J. W., Wayne-Current, K., Magnuson, W. G., and Pocha, A. Introducing gate arrays to engineers. *J. Semicustom ICs*, 1(2):34–41, 1983.

95. Anderson, R. C. *Multi-Source: Integrated Circuit Design and Procurement System.* R. C. Anderson, Inc., Los Altos, CA, 1st issue 1982.
96. Farina, D. Review of the above, VLSI Design. 3(4):39, 1982.
97. ICE Report. *Hybrid Microelectronics for the 80's.* Integrated Circuit Engineering Corporation, Scottsdale, AZ, 1983.
98. Iscoff, R. VHSIC/VLSI packaging update. Semiconductor International, 6(7):52−57, 1983.
99. Parris, S. R., and Nelson J. A. Practical considerations in VLSI packaging. *VLSI Design,* 3(6):45−49, 1982.
100. Various papers. Proc. Internepcon., Brighton, UK, 1983.
101. Fishman, D., and Cooper, N. Mounting leadless chip carriers onto printed circuit boards. *Hybrid Circuits,* 1(1):38−43, 1982.
102. Report. IC packaging and equipment interconnect practice. BPA Consultants, Dorking, UK, 1982.
103. Nicholson, B. Surface-mount technologies expand, but widespread use delayed. *EDN,* 28(26):224−231, 1983.
104. Fehr, G. K. Logic array packaging. LSI Logic Corporation Application Note A33, Milpitas, CA., 1982.
105. King, G. R. Will plastic packaging for semi-custom IC's be reliable enough for tomorrow's systems. *Proc. 3rd Int. Conf. on Semi-Custom ICs,* paper 2/22, London, 1983.
106. Simpson, N. Integrated circuit packaging trends. *Proc. 3rd Int. Conf. on Semi-Custom ICs,* paper 2/23, London, 1983.
107. Burggraaf, P. S. IC lead finishing: issues and options. Semiconductor International, 6(7):64−69, 1983.
108. Walker, R., and King, D. Testing Logic arrays. LSI Logic Corporation, Application Note A32, Milpitas, CA, 1982.

5
Custom-Specific Design Using Cell Libraries

Having considered the varied aspects of uncommitted masterslice products, which basically require only the final interconnection level(s) of the chip to be designed and fabricated, let us turn to the cell-library approach. As will be seen, this approach is likely to gain a greater share of the total custom-specific marketplace, but will not entirely supplant other methods.

5.1 BASIC CONCEPTS: POLYCELLS

In general, the cell-library approach involves no wafer diffusion stages before fabrication of the custom product; a full mask set is, therefore, necessary, exactly as with a fully handcrafted custom design. However, as we have previously noted, the cell-library approach capitalizes on the availability of a comprehensive range of functional elements, ranging, for example, from simple 2-input logic gates through to full adders, shift registers, and so on, these macros being fully designed and documented in the system library. The general advantages of the cell-library approach compared with masterslice array are (1) overall chip size for a given duty is smaller, since only the required circuit elements are present on the final chip, and (2) possibly increased circuit performance for a given state-of-art technology, due to better on-chip packing density. The size advantage is illustrated in Table 5.1.

TABLE 5.1 Typical Size of Chip Versus Design Methodology

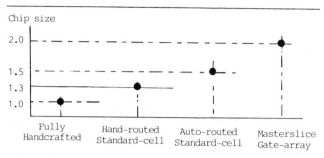

The general procedure for a cell-library-based design, therefore, is indicated in Table 5.2. Details of the simulation and functional check of the design against the original system specification will vary between different design systems, but must be present to close the loop between the design stages and the system requirements.

TABLE 5.2 The Basic Steps in Cell-Library Custom-Specific Design

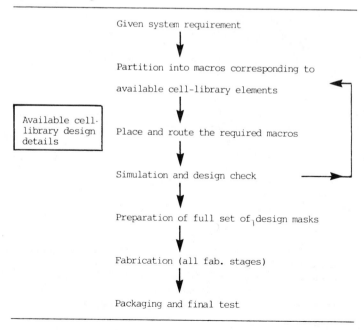

Because the complete floor plan layout of the final chip is being designed, there are no predetermined width wiring channels with a given maximum wiring capacity, as in the masterslice approach. Hence, the problems of fixed wiring channel capacities which were considered in the previous chapter do not arise, and there is freedom to make the interconnection wiring area to the exact requirements. However, when placing the standard cells on the floor plan layout it is still usual to maintain a column (row) placement in order to retain a structured interconnection topology. Figure 5.1 illustrates the general floor plan of a cell-library-based design (see also Chap. 1 Fig. 1.4).

In addition to the usual assembly of cells into columns or rows, the cells themselves are normally designed so that all have a consistent width but variable length, with the cell terminals spaced on a rigid grid spacing. This is illustrated in Figure 5.2. This philosophy is often termed the "polycell" approach, to distinguish between this situation and other occasions when cells may not retain a common dimension between different members of a library. It is clear that the polycell approach allows columns (rows) of abutting cells to be built up similar to the columns (rows) of fixed cells in a masterslice array, with the possibility of straight dc supply rails running through the assembly. However, there is also the freedom during the placement and routing procedure to randomly break the columns (rows) of polycells in order to provide routing through from one wiring channel to the next, this

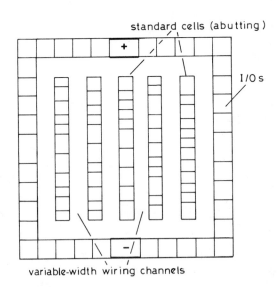

FIGURE 5.1   The floor plan of a typical cell-library design, with variable width wiring channels.

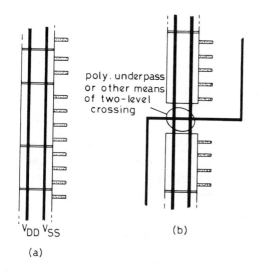

poly. underpass
or other means
of two-level
crossing

V_{DD} V_{SS}

(a)

(b)

FIGURE 5.2   The standard-grid (polycell) approach: (a) abutting
polycells with through dc supply lines, (b) break in abutment for
escape intercolumn routing.

flexibility, of course, not being possible with the fixed masterslice
approach. Such dog-legging from one wiring channel to another places
a greater demand upon the sophistication of the computer-aided design
(CAD) routing, or a requirement for hand intervention in the placement
and routing procedure, and therefore, may not be available.

It may also be appreciated from Figure 5.2 that the exact means of
connection between the metalization level interconnect routing and the
polycell terminals involves precisely the same fundamental considerations
as in the masterslice fixed-cell situation, and considerations such as
single-sided access, double-sided access and through-cell connections
which were discussed in detail in the previous chapter are equally relevant. However, with the cell-library approach there is the freedom to
design the polysilicon level connections from the cells into the wiring
channels as required without being constrained to some fixed polysilicon
level interconnect strategy as illustrated in Figures 4.22 to 4.24. Cyclic routing constraints, such as illustrated in Figure 4.19, may therefore be circumvented by appropriate floor plan management.

Many commercial cell-library designs provide both-side access to/
from the polycells, which implies a single-column (single-row) floor
plan, and therefore, the overall chip design may superficially look
similar to a masterslice array, but with much greater freedom at the
detailed design level. CAD software for routing may be very similar
between the two techniques.

All these considerations so far imply that the macros in the cell-library are small-scale/medium-scale integration (SSI/MSI) size macros, rather than larger entities such as complete programmable logic array (PLA) structures, blocks of memory, and the like. It is evident that when the size of available "standard parts" in the library reaches a certain level, the concept of columns or rows of standard-width poly-cells breaks down, and the chip design now involves the placement of large functional blocks (sometimes termed "supercells" or "megacells") in some structured relationship. This is indicated in Figure 5.3. Indeed, we are now entering the area of structured very-large-scale integration (VLSI) design, where handcrafting of a complete new chip down to individual device level becomes completely impractical due to the sheer volume of detail involved. Partitioning into large functional blocks and subsystems is the only way to handle such complexity, and hence we may consider that all VLSI design will involve a hierarchial design approach. We will refer to these considerations again in Section 5.7 [1–5].

At the other end of the cell-library scene, we find a further blurring of distinction, this time between masterslice arrays and the polycell approach. This is particularly so in complementary metal-oxide semiconductor (CMOS) technology. It will be recalled from Chapter 4 that in CMOS masterslice arrays a favored arrangement is a longitudinal assembly of *p*-channel and *n*-channel transistor pairs, as for example illustrated in Figures 4.6 and 4.23. The interconnection commitment of these transistor-pairs into functional primitives has also been described. Such commitments may of course be used as standard-cell (polycell) designs, and hence it is possible that very similar functional primitives and macros may be present in both masterslice and cell-library CMOS realizations, both basically being transistor-pair assemblies. The distinction which remains, however, is that the masterslice realization

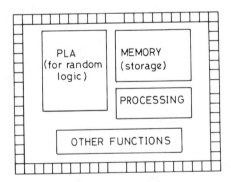

FIGURE 5.3   The extension to the cell-library approach to VLSI, using large building blocks from the library.

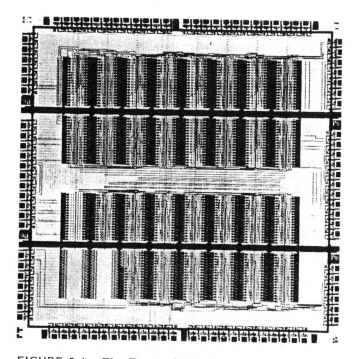

FIGURE 5.4 The Ferranti AR series gate array, with provision for automatic redesign of the floor plan layout. (Photo courtesy Ferranti Electronics, Oldham, England.)

involves only the final interconnection mask set with a fixed floor plan topology, whereas the full mask set together with floor plan flexibility is present in the cell-library product.

There are yet further facets to blur this boundary area. Figure 5.4 shows the layout of the Ferranti AR series gate array, which is a CDI technology masterslice array with a silicon-compiler type CAD software suite for its dedication. However, the CAD software package generates all mask levels for each custom design; if the particular design will not fit on an existing gate-array floor plan, due to high wiring channel interconnection densities or positional input-output (I/O) constraints, then a new size array is created on a customized floor plan, and a new full mask set is employed for the chip production. The advantage claimed for this approach is to guarantee 100% automatic placement and routing (silicon compilation) for all designs, with full flexibility down to the device level of design [6].

Another variation which spans the rigid masterslice/cell-library boundary is the recently announced CMOS cell array (CCA) product

TABLE 5.3   The General Distribution of Cell-Library Based Custom-Design, with Blurred Boundaries Between Adjacent Techniques

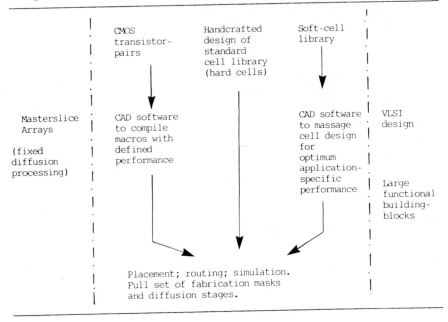

of Bell Labs [7–10]. This approach uses a library of functions in the form of pre-characterized cells, the cells being assembled from p- and n-type transistor-pairs as previously considered. However, instead of providing field-oxide isolation between macros, isolation is achieved by using transistor-pairs, isolation being achieved by connecting the gate of the p-type device to V_{DD} and the gate of the n-type device to V_{SS}, thus forming permanently off devices between the active macros. Since the layout of the continuous transistor-pairs is common for all final commitments, cell boundaries being defined halfway through the processing stages at the polysilicon level of fabrication, it is now possible to preprocess and stockpile wafers for subsequent commitment up to, but not including, the polysilicon level. Wafer dedication, therefore, involves more stages than a conventional masterslice array, since it involves the polysilicon stage onwards, but not the full mask set which is the normal standard-cell requirement.

Finally we should note that there is an increasing interest in the ability to modify the so-called standard cells in a cell-library system, so as to optimize the chip performance for specific duties. This may involve shrinking device sizes in given cells to increase speed, or increasing sizes to improve drive capabilities, or removal of unwanted cell inputs or outputs. We will refer to such "soft-cell" considerations in Section 5.7. Table 5.3 attempts to summarize this still evolving scenario. For futher information see many available publications [7—18].

5.2 PLACEMENT CONSIDERATIONS

To a certain extent it may be argued that functional placement in a standard-cell custom-specific design is not as critical as in masterslice array dedication, since wiring channels can be made as wide as necessary to accommodate the intercell routing requirements. Hence, there is no fixed wiring track capacity which cannot be exceeded. On the other hand, it may be argued that as long as it is possible to route a masterslice array within the available wiring track capacity, there is no need to search for better functional placement which would reduce the maximum track density, since this would not reduce the fixed masterslice chip size, whereas with a cell-library design a reduction in maximum track density will allow a corresponding reduction in silicon area.

The latter argument is somewhat more realistic, and there is a noticeable desire for good placement in cell-library realizations in order to eliminate peaks in track density routings, and to achieve maximum on-chip performance. This is not to say that placement on a masterslice array is unimportant, since optimal placement may enhance the final chip electrical performance.

A search of the generally available literature will reveal wide mention of (1) device modelling, (2) circuit simulation and verification, (3) placement and routing, (4) layout rule checking, and (5) testvector generation, in connection with custom specific microelectronics, these being the principle areas of relevance when considering CAD. It is significant that placement is often the Cinderella of this scenario, with much wider detailed publication of the other areas and their software availability being covered. For example "placement" is not indexed as such in many otherwise comprehensive texts [1,3,4]. Indeed, the current situation very largely is that placement (or rather re-placement) remains at hand-intervention level from initial primitive and macro placement decisions, the initial placement being possibly based upon a linear macro listing from which a "folded-snake" first placement is obtained [19—23]. The graphics CAD software for hand intervention may have facilities for: (1) rotating cells, (2) specifying mirror-image cell layouts, (3) interchanging cells, (4) interchanging columns (rows) of cells and possibly other re-placement facilities. The use of such interactive CAD facilities relies upon the unique capabilities of the human operator backed up by the capabilities of the routing software to reroute and display the consequences of re-placement very

rapidly. The latter may be a local rerouting confined to the area of
the layout affected by the re-placement, or global if the effects are distributed.

5.2.1 Placement Considerations for Polycell Layouts

The available literature on placement techniques originates in the 1960s,
concerned initially with the placement of circuits on printed wiring
boards and other subassemblies in order to minimize module interconnection density and optimize the modularity for maintenance and repair
purposes. In general, three classes of placement techniques may be
observed in this work:

1. An iterative replacement procedure, in which circuit elements
    (primitives or macros) are interchanged, possibly at random or
    perhaps under some connectivity indicator, the result of each
    action being observed.
2. Constructive initial placement, in which an analysis of the de-
    gree of connectivity between circuits is first made, from which
    a placement solution is generated.
3. Branch-and-bound methods, whereby circuits are initially divid-
    ed into large groupings, each of which is then repeatedly sub-
    divided to form a decision tree of associated module groupings.
    Interchanges of the branches of this tree may also be performed.

The first technique is a purely iterative process, which can be terminated at any stage to give a functionally complete placement, but the
latter two are constructive, producing a placement solution only on
final completion. It is possible to combine these methods to some extent, for example, to perform an iterative replacement of modules after
some initial constructive placement. For a comprehensive survey of
the work, see Hanan and Kurtzberg [24].

These early considerations had a number of practical constraints to
take into consideration, such as the number of pins per printed wiring
board (PWB) connector or the space on a PWB for the wiring tracks,
all of which clearly have their parallel in custom and semicustom IC design. Since it is generally impossible to satisfy all criteria simultaneously, and since many are specific to a given product (cf.
the available wiring channel capacity on a particular masterslice array), a compromise single parameter which represents a norm for
optimal placement was adopted. This parameter was the *total interconnection length* between all the modules of the system, which should
be minimized. However, since some connections in a system may be
more important than others, individual connections may be given a
weighting (importance) value, optimal placement now being the search
for the *minimum weighted interconnection length* summation.

The interconnect nets, however, may in the unrestricted case have different possible topologies, such as shown in Figure 5.5. Not all of these carry over precisely to the masterslice or polycell area, and indeed we may define a more specific interconnection "length" for polycell and masterslice arrays than these more general topologies. However, the concept of some connection length between primitives or macros remains.

Most of this earlier work outside the custom IC design field remains the basic placement theory. There has been relatively little additional fundamental development, but rather we have seen the adaptation of basic concepts to the specific problems of IC placement.

With floor plan layouts such as shown in Figure 5.1, where standard width (height) polycells are assembled in columns (rows) with interspersed wiring channels of appropriate capacity, the concept of minimization of the total intercell wiring length per chip is common. However, with interconnections restricted to horizontal and vertical paths, there is what is termed a "Manhattan interconnection topology," with the Manhattan distance M between two points x_1, y_1, x_2, y_2 being defined by*

$$M = \left\{ |x_1 - x_2| + |y_1 - y_2| \right\}$$

The direct distance between these points, namely

$$\left\{ (x_1 - x_2)^2 + (y_1 - y_2)^2 \right\}^{\frac{1}{2}}$$

has no significance. Alternative definitions of length which may be used are

1. The summation of distances along the wiring channels only, the "trunk length," ignoring individual cell connections into the wiring channels and all perimeter connections [25].
2. One-half of the perimeter of the smallest rectangle which encompasses all terminal points of a complete net.

Note that the latter definition is not necessarily the same as the summations of the lengths of the individual connections which go to make up a complete net, but is equivalent to the lowest bound tree length in orthogonal routing between all points of a net, provided the net routing

---

*See, for example, the orthogonal street plan of Manhattan, where taxi distances are based upon the number of north-south and east-west blocks between two points.

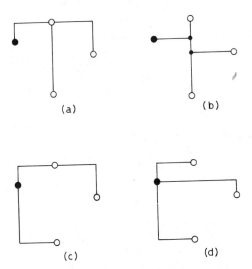

FIGURE 5.5   Interconnection nets with different interconnect philosophies, source cell indicated by solid circle, driven cells by open circle.
(a) Minimum spanning tree, length 20 units, (b) Steiner tree, length
15 units, (c) chain, length 21 units, (d) individual source to driven
cell connections, 24 units.

is not blocked by intervening obstacles. In all cases, weighting of
connection distances may be made in order to emphasize critical connections at the expense of less important connections.
      Hence, if a chip consists of I total connections, then using for ex-
ample the Manhattan length, the total on-chip Manhattan length L_M
is given by

$$L_M = \sum_{i=1}^{I} M_i$$

and if each connection is given a weighting factor w, then the total
weighted Manhattan length is

$$L_{WM} = \sum_{i=1}^{I} w_i M_i$$

Minimization of this parameter may be considered as a goal for optimal
placement.
      Other "goodness" parameters which have been suggested include
minimization of the total number of interconnect crossovers [26], and

minimization of the total number of 90° turns in the completed routing [27], assuming north-south/east-west routing directions only in the latter. However, all these are possible figures of merit for comparing the resultant routing of one placement with another; they do not per se provide a means of determining a "best" placement.

The actual placement algorithms which may be used for polycell structures include linear placement, clustering, mincut placement, and variants on these themes. In linear placement, it is assumed that all cells are first arranged in a single horizontal row. In theory it would be possible to consider all possible permutations of the order of the cells in the quest for minimum total Manhattan or other distance summation, but with T cells this would involve a total of $\{\frac{1}{2}T!\}$ permutations, excluding mirror image, which is far too large to consider for any practical situation. Hence, some clustering algorithm is desirable in order to provide a first global placement, following which local perturbation of cells or cell clusters may be invoked to fine-tune the final placement.

Clustering aims to bring together strongly associated cells at the expense of dispersing those which have little commonality. The connectivity V_{PQ} between two cells P and Q may be defined as

$$V_{PQ} = \left\{ \frac{f(P)C_{PQ}}{D_P} + \frac{f(Q)C_{PQ}}{D_Q} \right\}$$

where $f(P)$, $f(Q)$ are empirical factors related to the size of cells P, Q respectively,

C_{PQ} = number of connections between P and Q

D_P = number of connections from P not routed to Q

D_Q = number of connections from Q not routed to P

For example, in Figure 5.6 taking $f(P)$, $F(Q)$, $f(R)$ and $f(S)$ all as unity, we have:

$$V_{PQ} = \left\{ \frac{4}{1} + \frac{4}{3} \right\} = 5.33$$

$$V_{QR} = \left\{ \frac{1}{6} + \frac{1}{5} \right\} = 0.37$$

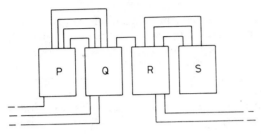

FIGURE 5.6   Clustering of cells and connectivity parameters.

$$V_{RS} = \left\{\frac{3}{6} + \frac{3}{0}\right\} = \infty$$

$$V_{PR} = V_{PS} = V_{QS} = 0$$

From this it will be noted that loosely connected cells have a low connectivity value, being zero if no interconnections exist, while any cell which is 100% connected to another has infinite connectivity value.

In this example it is obvious that cell S should be adjacent to cell R. Equally there is no requirement for cells P and R, cells P and S, and cells Q and S to be specifically paired with respect to each other. The clustering algorithm therefore proceeds as follows:

1. Compute the connectivity value of all pairs of cells
2. Combine the two cells with the highest connectivity factor; call this a new cell
3. Recalculate all cell connectivity factors
4. Repeat (2) and (3) until all cells have been clustered

This is the outline procedure. Problems however arise as "cells" become large in this clustering procedure, since they tend to dominate the connectivity values. This is where the introduction of the f(P), f(Q), . . . factors becomes necessary, in order to counteract the increasingly dominating effects of the large macroclusters. Other difficulties may arise due to connections running between three or more cells, which may require further empirical adjustments. However, a reasonably sensible clustering of the original cells results, but whether it differs markedly from an initial placement which represents a logical layout of the given circuit schematic is possibly debatable.

Having produced some initial linear cell clustering, perturbation of the cell clusters can be undertaken. Since we are now perturbing the positions of clusters, exhaustive perturbation may be possible; on the other hand, perturbation of cluster positions may not have a very strong effect upon the total interconnect length or other figure of merit being used, and therefore, optimum placement can never be guaranteed. Finally the single row of placed cells requires to be folded into rows (columns) to produce a chip layout of roughly equal north-south and east-west dimensions.

The alternative mincut placement approach is possibly not so directly relevant to masterslice arrays and polycell layouts as the previous techniques and their variants, since it does not commence with a one-dimension assembly of cells. Instead it allows some initial, possibly entirely random, two-dimensional placement, such as a sea of cells or equivalent, as illustrated in Figure 5.7a. A cut C1 is applied to divide the assembly into two blocks B1 and B2, the interconnect lines crossing this cut being termed "single cuts." The mincut placement algorithm then seeks to minimize the number of signal cuts by swapping cells between B1 and B2. This procedure is then repeated by the introduction of a second cut (see Fig. 5.7b), with the minimization of signal cuts across C2 then being sought. Further cuts and repeats of the minimum signal-cut search continue until a satisfactory solution has been reached, which may be at the ultimate individual cell per block level or terminated at some preceding stage.

This mincut procedure is an application of a previous partitioning algorithm known as the Kernighan-Lin partitioning algorithm [28]. However, applied to the polycell placement problem there is the practical situation that the physical size of cells being swapped across the mincut boundaries may not be the same, and hence a final chip layout

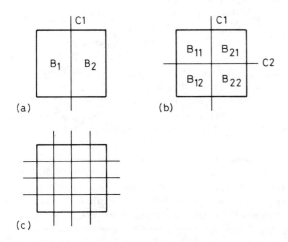

FIGURE 5.7 Mincut placement procedure: (a) first cut C1, (b) second cut C2, (c) subsequent cuts.

to implement placement based only upon the minimization of signal cuts may not be satisfactory.

Other placement algorithms include:

1. Iterative pairwise interchange, in which pairs of modules are interchanged; if a reduced total interconnect length results, then the interchange is accepted; if not the original placement is retained.
2. Relaxation methods, in which each connection is thought of as a stretched elastic string containing a force f proportional to the length l of the connection; relaxation techniques to minimize the total force Σf between all cells by re-placement are then applied.
3. Force-directed pairwise interchange, which is a combination of 1 and 2 above.
4. The use of "seed cells," that is, an aribtrary cell is chosen and all closely connected cells to this seed cell are then drawn into proximity; further seed cells are chosen until all cells are finally grouped.

In practice it is usually found that some combination of techniques is proposed giving, for example, a first global placement followed by local improvement re-placement to achieve a final acceptable solution. However, it must be borne in mind that it still remains necessary to route the cells or macros (see Sec. 5.3) and this may override the result of a purely placement algorithm. Routing may therefore dictate the final placement. As a concluding illustration of a commercial approach to placement which incorporated multi-algorithms and single or two-dimensional placement, the Bell Laboratories PLAC program may be mentioned [29]. In the PLAC program, an initial placement phase is undertaken, which sequentially selects unplaced cells according to their connectivity to chosen seeded cells, and builds up blocks so as to minimize the total net routing length. Routing length per net is taken as the half-perimeter distances (see previously). A second iterative improvement phase then performs pairwise exchanges of cells to reduce the total routing length. However, to allow for unequal size cells, the initial placement and the iterative improvement phases are interlaced so that iterative improvements can be attempted as the initial placement is being built up, rather than leave it until the initial placement of all cells is completed. An important feature, however, is the interactive capability with the designer, who communicates his expertise to the placement problem by the selection and preplacement of key seed cells, this being particularly relevant where fixed floor plan topologies such as masterslice or polycell layouts are involved.

For further published information and discussion of overall concepts, see Hanan and Kertzberg and others [24,30,31,117]; for mincut

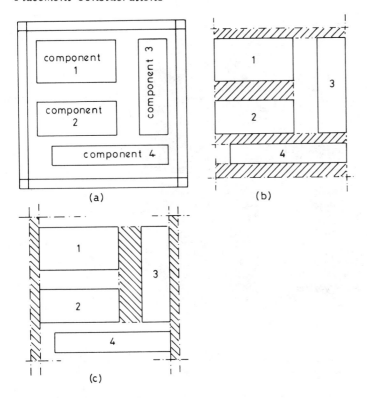

FIGURE 5.8 Placement of large macros or predesigned subsystems:
(a) global placement, (b) the definition of horizontal wiring channels,
(c) the definition of vertical wiring channels.

placement see Breur and others [32–34]; for iterative re-placement and
pairwise-interchange see Steinberg and others [35,36]; for force-
directed placement algorithms see Wilson and Smith and others [30,37,
38], and finally for specific technology discussions see Shiraishi and
Hirose and others [39–43]. For fundamental mathematical developments
underlying many of the algorithms, reference may be made to Berge and
other standard texts [44–46]. Finally, the execution of placement algorithms on array processors rather than on conventional von Neumann
computers has been considered by Chyan and Breur [47].

5.5.2 Placement Considerations for Large Macro Layouts

The placement of non-polycell layouts relaxes the topology of standard
width (height) cells, and generally precludes the consideration of an

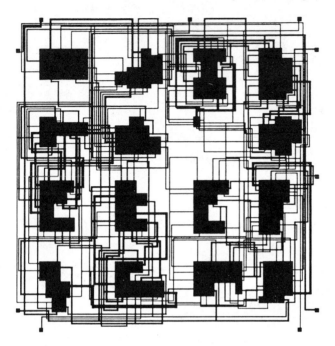

FIGURE 5.9   A placement and routing example with irregular shaped cells. (From Ref. 50.)

initial placement in a single string. Instead we have a two-dimensional placement problem, which may involve the placement and subsequent routing of rectangular macros having arbitrary areas and length/width ratios. We are now probably more into very-large-scale integrated (VLSI) design, with the on-chip assembly of relatively few but large predesigned subassemblies such as programmable logic arrays (PLAs), memory, and so on. Figure 5.8 illustrates this situation.

To a certain degree, if relatively few but large macros are present, the problem is simpler than the polycell placement problem. The final chip layout should be rectangular, ideally approximately square in order to be conveniently carried in conventional integrated circuit (IC) mounting packages (see Fig. 4.5 for example), and hence there may be relatively few ways to arrange the placement without incurring a large wasted silicon area penalty. The positioning of input and output terminals per macro and the required interconnect routing may also dictate a particular floor plan without a great deal of flexibility.

The placement of large macros for custom VLSI has been considered by Supowit and Slutz [48] and others. It will be found that wiring channel requirements and routing areas play an intimate part in the placement considerations, the two aspects of placement and routing no longer being possible in isolation. The goal of placement may now be stated as: *the minimization of the area of the minimum rectangle bounding both the macros and their interconnection routing.* The routing requirements are frequently incorporated by the separate consideration of horizontal channels and vertical channels [49] (see Figs. 5.8b and c), which allows the floor plan of the complete chip to be more easily optimized. However, unless the large macros themselves are able to be adjusted in some way, it remains debatable whether any placement algorithm for few, large rectangular macros will give a better placement than the original equipment manufacturer (OEM) can himself allocate, based upon his global appreciation of the system being assembled onchip. However, if irregularly shaped macros or other structural variations are present, then CAD placement becomes increasingly significant. Figure 5.9 illustrates a previously published placement result with irregular shaped macros [50].

For additional information on VLSI placement and irregular shaped cells, see further available publications [25,48—53].

5.3 ROUTING CONSIDERATIONS

The algorithms which can be found employed for the routing of masterslice and polycell assemblies likewise have their foundations in routing developments for printed wiring boards (PWBs) and similar applications. Routing on silicon, however, may have more severe constraints, such as single-layer-metal interconnection only or limited wiring channel capacities. Nevertheless the fundamentals of the problem remain the same.

The basic steps which have to be followed in a general routing procedure are:

1. Preparation of the interconnection net lists.
2. Decisions (if necessary) concerning what connections shall be on which layer of the interconnection structure.
3. Decisions (if necessary) concerning the order in which connections shall be routed, together with any critical signal paths.
4. The final routing layout for each interconnection net.

Items 2 and 3 above may be less significant for IC routing than for, say, complex multilayer PWB design, but with silicon circuit density still increasing the need for more sophisticated multilayer on-chip interconnect may increase the significance of these aspects. However, we will not specifically need to investigate them herewith.

The generation of the interconnection net list may involve only a simple listing of cell points which have to be connected, with no regard to order or physical location, or it may be a more sophisticated listing taking these factors into account. A considerable complexity can accrue if the net list has to contain topological information, and to some extent such a list may contain the final routing instructions. Details may be found discussed in Akers and elsewhere [54,55]. For the majority of present masterslice and polycell applications, simple net lists such as provided by front-end entry CAD (see Chap. 4, Fig. 4.27, etc.) are appropriate. This leaves us with the consideration of the final routing of the net lists.

In the routing of on-chip interconnections, path directions are usu-ally restricted to north-south and east-west single-layer interconnec-tions, and in addition possibly confined to precise intercolumn wiring channels. Hence, Manhattan paths and distances, as introduced in the preceding section, are present. *General routers* with one interconnec-tion level plus crossunders are applicable for such structures, or *chan-nel routers* where defined wiring channels between columns (rows) of cells are present.

5.3.1 General Routers

General routers attempt to find the shortest Manhattan path between two adjacent points in a net, around any intervening obstructions. The two most cited techniques are (1) Lee's routing algorithm, and (2) Hightower's routing algorithm. In Lee's algorithm [56], sometimes referred to as a "wavefront" or a "maze-running" router, the path to be routed follows a rigid rectangular grid construction, moving from one cell in the grid to an immediately adjacent cell. For example, con-sider the situation illustrated in Figure 5.10a, where the white cells indicate space available for interconnection routing, and the black cells are obstructions due to prior routing or other items. If we require to route from source cell position S to target cell position T, we first num-ber the available white cells as follows:

1. All cells immediately surrounding S are number "1"
2. All cells immediately surrounding the cells numbered "1" are numbered "2"
3. Continue numbering cells surrounding every already numbered cell with the next higher integer number until the target T is encountered

This is shown in Figure 5.10b.

The Lee's routing algorithm then selects a routing path between S and T by backtracking from T towards S, always moving from one

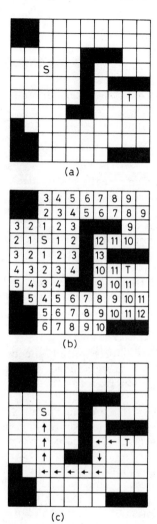

(a)

(b)

(c)

FIGURE 5.10 Lee's algorithm for single-level routing: (a) grid division of the floor plan, with S the source mode and T the target mode, (b) grid numbering spreading out from S towards T, (c) possible routing back from T to S.

grid square to a next lower-numbered cell, such as shown in Figure 5.10c. The advantages of Lee's algorithm are:

1. Provided the initial cell numbering can be extended from S to T, then a path between S and T is guaranteed
2. The path chosen will always be the shortest Manhattan distance between S and T
3. It can readily cope with nets with more than two points S and T

Disadvantages, however, include:

1. Complications occur where there is a choice of cells in the back-tracking, since in theory any cell numbered $(x - 1)$ may be chosen to follow cell x
2. The choice of a particular route between S and T may preclude subsequent net list connections from being routed, with deletion and rerouting (by hand) of previously completed nets being necessary for 100% completion
3. A considerable and unnecessary area of the available wiring space may be covered and enumerated by the grid radiating outwards from S, with high computer memory requirements and long central processing unit running time

Various techniques to refine the basic procedure have been advanced [54,57,58], particularly to avoid the labelling of matrix squares which are generally running in an inappropriate direction from the target cell T.

Hightower's general routing algorithm [59] is also a maze-running algorithm, but unlike Lee's method does not require the comprehensive numbering of grid squares radiating from S towards T. Instead we establish north-south and east-west lines starting from S, along which we search for the nearest point from which a perpendicular escape route towards T avoiding constraints can be made. This is a new source point S_1. A similar procedure is started at T, aiming generally toward S, giving a new source point T_1. This is illustrated in Figure 5.11a.

This procedure is repeated from the new source and target points S_1 and T_1, until eventually a line which traces back to S cuts a line which traces back to T. Hence, a Manhattan connection path between S and T is established. This is shown in Figure 5.11b. The advantages of this and similar algorithms include: (1) the algorithm tends to find the path with the minimum number of changes in direction; (2) reduced storage and computing time compared with Lee's algorithm; and (3) the final routing path is likely to have less likelihood to block subsequent routing paths. Against these points, the minimum Manhattan distance is not necessarily found, and, like Lee's method, no guarantee of 100% automatic routing is available. Modifications to improve the algorithm have been reported [60].

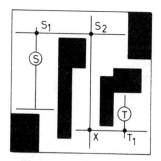

FIGURE 5.11 Hightower's algorithm for single-level routing: (a) initial routes from S and T, (b) final routing, connection path $SS_1S_2XT_1T$, X being the meeting point in the routing algorithm.

5.3.2 Channel Routers

When columns (rows) of cells with intercolumn wiring channels are present, more specific routing algorithms may be employed. Furthermore, rather than consider the global routing problem for each net in the net list, a hierarchical routing procedure may be possible, wherein for example, the routing of each intercolumn wiring channel may be undertaken, followed by the perimeter wiring involving end channel and input/output (I/O) connections. Such a "divide-and-conquer" approach considerably eases the overall complexity of routing procedures, since the global requirements are tackled in smaller sections one at a time. The floor plan design of the chip plays a vital role in these possibilities. Figure 4.24 (Chap. 4) illustrated a possible philosophy for the divorcing of wiring channel routing from perimeter routing, albeit at the expense of second-level polysilicon interconnect; details of this philosophy which may be extended to polycells and two-layer metal may be found published [23].

With channel routers, we again normally assume that the channel may be represented as a rectangular grid with uniform spacing. Assuming that the rows of cells and wiring channels are horizontal, then the cell terminals are spaced at uniform grid positions along the wiring channel, with the connections to be routed running as horizontal paths along the channel, connecting with vertical polysilicon (or other lower-level interconnection medium) fingers from the cell terminals.

The *left-edge channel router* of Hashimoto and Stevens [61] is well known. Its development is shown in Figure 5.12. It starts by selecting the extreme left-hand cell connection and routes this at the top of the channel. In the case of two directly facing cell connections, it first selects the top one of the two. The next connection routed is the first cell connection to the right of the routing just completed which is able to be routed at the same track level in the wiring channel as the first

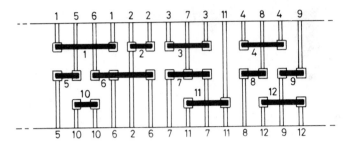

FIGURE 5.12   The left edge channel router, sequentially routing nets
1,2,3. . ., with no cyclic constraints present.

connection. This procedure repeats until no more connections can be
routed at this level, following which it returns to the left-hand edge
and selects the next available pin for routing at the next level track
position in the channel.

It will be observed that this left-edge channel routing algorithm is
subject to two fundamental constraints, namely (1) a horizontal constraint, in that a horizontal overlap of nets on the same track level cannot be allowed and (2) a vertical constraint, in that a vertical overlap
of directly facing connections from cells cannot be allowed. When these
two constraints are combined, then we may encounter a cyclic constraint
which precludes routing a required net. This feature has already been
discussed in Chapter 4 (see Fig. 4.19 in particular*), and may be overcome by appropriate management of the positions of the facing cell terminals, or by re-placement procedures [61−63].

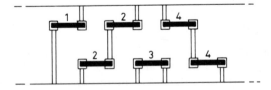

FIGURE 5.13   The dog-leg router of Deutsch, which may circumvent
cyclic constraints.

---

*The majority of illustrations in Chapter 4 were shown with vertical
wiring channels; however the terminology "left-edge channel router"
requires that we reorientate ourselves to horizontal channels for the
present discussions.

TABLE 5.4 Results of Various Channel-Routing Algorithms Applied to Various Benchmark Net Lists

Benchmark net list		Track density given by various routing algorithms				
Example	Minimum lower bound track density	Left edge	Zonal	Greedy	BARS	Hierarchical
1	12	14	12	—	—	—
3(a)	15	18	15	—	—	—
3(b)	17	20	17	—	—	—
3(c)	18	19	18	—	—	—
4(b)	17	23	17	—	—	—
5	20	22	20	—	—	—
"Difficult"	19	21	20	20	19	19

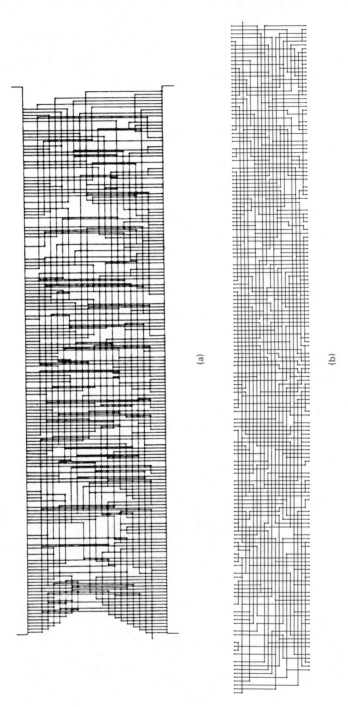

(a)

(b)

FIGURE 5.14   The channel routing of Deutsch's "Difficult" net list by two different channel routers (a) BARS (b) Hierarchical.

TABLE 5.5 Present Relationships Between Gate Count and CPU Time

Chip complexity	CPU time t		
	Channel routing	Maze routing	Test generation
1,000 gates:	1 minute	1 minute	1 minute
10,000 gates:	16 minutes	106 hours	16 hours
100,000 gates:	126 minutes	144 hours	1.6 years
Relationships:	$t \alpha \; \{\text{gates}\}^{1.2}$	$t \alpha \; \{\text{gates}\}^{2}$	$t \alpha \; \{\text{gates}\}^{3}$

Source: Details courtesy Pyne, A., Trends in semicustom integrated circuits—an overview. *Microelectronics J.*, 14(3):10, 1983.

The inclusion of dog-legs in the channel router may also overcome cyclic constraints, but at the expense of a more complex routing procedure. The dog-leg router of Deutsch [64] allows channel routing to be completed by means such as shown in Figure 5.13. Branch-and-bound techniques are also available [63], but may involve excessive computer run time if many iterations of the branching are investigated.

Other variants of channel routing algorithms may be found, including Zonal [65], BARS [23], Greedy [66], and Hierarchical [67]. A benchmark for these various algorithms has evolved over recent years, this being a series of net lists which the channel-routing algorithm is asked to route. The most complex of these net lists is Deutsch's "Difficult" example, which has a theoretical lower bound track density of 19 tracks across the wiring channel to fully route, but which until recently could not be routed by available channel-routing algorithms under 20 track widths per channel. Recent algorithms now realize the lower bound. Table 5.4 gives the present status; Figure 5.14 shows the "Difficult" benchmark example routed by two channel routing algorithms.

It will be appreciated that considerable development of algorithms for detailed channel routing has been undertaken. The global problem and the question of a possible global router which first determines which net lists shall be in which channels, however, remains a less well-defined area, involving as it may global placement as well as routing. In general, the global router should distribute the required interconnections as evenly as possible across the array, leaving the channel router to perform the detailed interconnections within each channel, possibly invoking cell-swapping or other means to optimize the track wiring density within each wiring channel or pairs of wiring channels. For additional discussions see Leong and Liu and others [68–70,117].

As a final point concerning the expediency of CAD routing, it will be evident that as chip size increases, routing time will also increase. Table 5.5 gives recently suggested central processing unit (CPU) times versus chip complexity; pressure for more efficient hierarchical placement and routing procedures to reduce the exponential values in these CPU times will become increasingly significant as custom design moves from LSI towards VLSI complexity.

5.4 OEM COMMITMENT PROCEDURES

The current situation regarding the OEM designer's involvement in the complete design cycle for cell-library-based custom circuits is that there is somewhat less scope for shared development than in the master-slice array case. In general the vendor will retain the intimate details of his cell layouts in order to retain commercial security, and hence only cell outline and parameter data may be made available. Should the OEM designer merely partition his circuit schematic into available functional primitives, the vendor then taking over for the detailed chip design and verification, OEM involvement is little different from that described in Chapter 4 where the vendor designs and produces a masterslice array product to the customer's requirements. Figure 5.15 illustrates such a customer interface for cell-library design.

If the OEM deliberately does not wish to become involved at all in the design cycle, then he may not even wish to choose between a masterslice array product or a cell-library-based product for his requirements. In this case there is the possibility of an independent design house becoming involved, which can offer fully independent design services. Such a situation is illustrated in Table 5.6.

However, should the OEM wish to participate in the design cycle, increasing involvement broadly as follows is possible:

1. Hand layout of chip, using vendor's skeleton outline data and performance details per cell; hard-copy diagram(s) back to vendor for checking and fabrication.
2. CAD layout of chip, using low-cost terminals running vendor's software containing skeleton outline data and performance details etc., magnetic tape of layout details back to vendor for checking and fabrication.
3. As (2), but with more comprehensive simulation and checking CAD software from either vendor or elsewhere (e.g., workstations running vendor's product).
4. Similar to (3), but using on-line vendor's CAD resources.
5. Full purchase or license of vendor's cell-library details; in-house comprehensive CAD resources for placement and routing, simulation, testing, preparation of pattern-generation magnetic tape.

TABLE 5.6 The Vendor-Independent Design House—Fabrication Route for Custom-Specific Circuits, with Cell-Library or Masterslice Array or Full-Custom Choices to the Customer

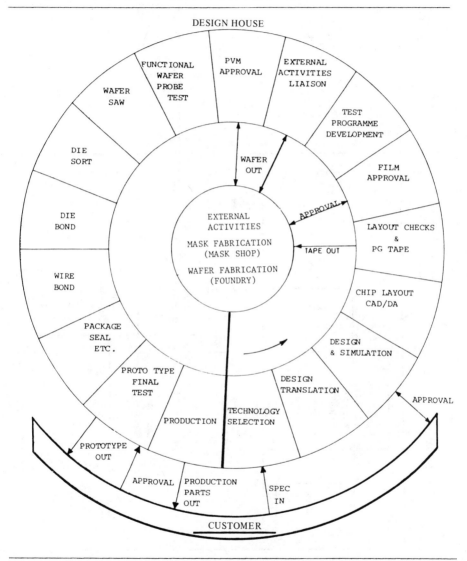

Source: Chart courtesy Silicon Microsystems, Malmesbury, UK.

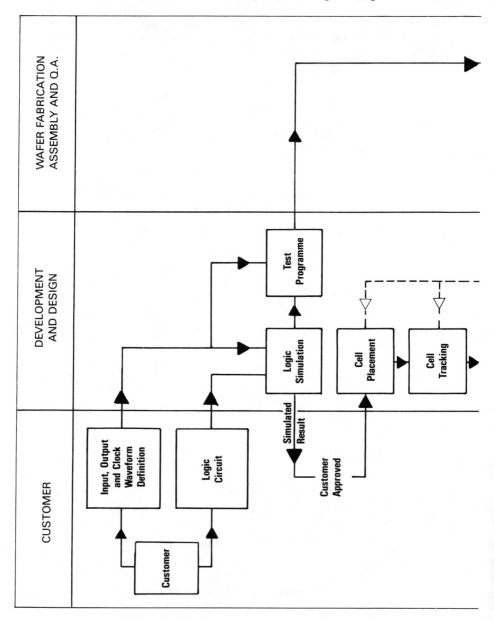

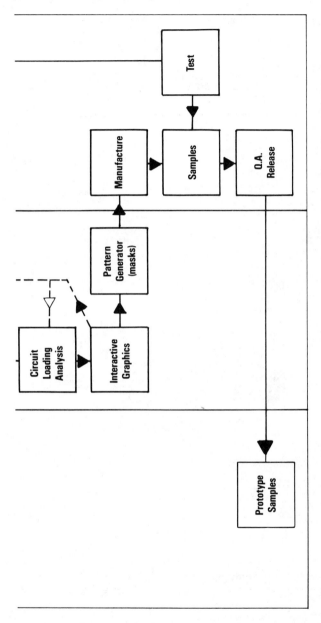

FIGURE 5.15 The custom/vendor interface for cell library custom-specific circuits. (Details courtesy Marconi Electronics Devices Ltd., Lincoln UK.)

TABLE 5.7 Customer-Vendor Interfaces Available with AMI Standard Cell Design[a]

	AMI Development (minimum customer involvement)	Shared Development	Customer-intensive Development
Functional specification	C	C	C
Logic diagram	C	C	C
Breadboard (if built)	C	C	C
Logic diagram partitioned into standard cells	O	C	C
Logic simulation	A	C	C
Circuit design	A	O	C
Cell placement	A	O	O
Cell routing	A	A	O
PG mag. tape	A	A	O
Mask fab.	A	A	A
Wafer fab.	A	A	A
Assembly	A	A	A
Acceptance test	A	A	A
Complete chip test	A	A	A

[a] A = AMI task; C = customer task; O = optional.
Source: Details based upon AMI Commercial Brochures, 1982.

FIGURE 5.16   Hand placement and routing of a cell-library chip, using skeleton outlines on a standard grid matrix.   (Details courtesy of Dumont Alphatron, Inc., Cupertino, CA.)

Thus variable interface positions are possible, depending upon the customer's CAD and other resources.   The choice of interface offered by different vendors may vary both in position and flexibility.   As an example Table 5.7 illustrates the possibilities listed by American Microsystems, Inc., which is typical of a flexible customer/vendor relationship, with the vendor retaining the final product fabrication.

Figure 5.16 illustrates the level of customer involvement at the design level with the minimum of sophistication.   This may be compared very closely with the hand-commitment of masterslice arrays illustrated in Chapter 4.   The higher levels of customer sophistication involving CAD tools may be visualized from further illustrations in Chapter 4, for example Figure 4.28 which has both masterslice and cell-library commitment facilities.

Further details of commercially available products, including CAD hardware and software, will be given in the following section.

5.5 COMMERCIAL STATE OF THE ART;
   COMMERCIAL AVAILABILITY

The commercial state of the art and the commercial availability of both
standard-cell products and CAD hardware and software is still in a
phase of rapid evolution, which makes discussion difficult in any comprehensive and lasting manner.

5.5.1 Semicustom Products

Reviewing first the silicon product side, we have considered in Sec. 4.8
of the previous chapter the number of industrial organizations active
in the semicustom silicon product area, and also the silicon technologies
which are represented. Not all of these companies and not all of the
several technologies, however, are present in the cell-library field.
In general, cell-library technologies are confined to (1) metal-gate
nMOS, (2) metal-gate CMOS, (3) silicon-gate nMOS, (4) silicon-gate
CMOS, (5) silicon-on-sapphire CMOS, with silicon-gate CMOS becoming
the increasingly accepted leader in the field. The current state of the
art is 5 micron geometry shortly scaling down to 3 or even 2 microns.
Silicon-on-sapphire CMOS is represented by RCA, long-time exponents
of this technology, and also by ASEA HAFO, with headquarters and
processing facilities in Sweden. Other large companies may also offer
insulated substrate technology in the future.
    The almost complete lack of bipolar technology in the cell-library
field is at first sight surprising, but the increasing dominance of CMOS
is largely due to the efficient manner in which both simple and complex
cells can be handcrafted in CMOS (see Chap. 4, Fig. 4.13 for example)
and the excellent speed-power product capability.
    Among the companies offering cell-library products for the OEM
market are the following:

AMCC	Master Logic
AMI	Matra
Array Technology	MCE Semiconductors
Harris Corporation	MEDL
Hughes	Mostek
Intel	Plessey
IMP	RCA

Signetics

Silicon Microsystems

Texas Instruments

VLSI Technology

This list is not guaranteed to be complete or currently correct, and inclusions and omissions are in no way indicative of the scope of activities in this area. Equally, some commercial products may be based upon transistor-pair macros which are common with uncommitted masterslice designs, rather than upon unique individually handcrafted layouts for each cell-library macro.

Unlike the masterslice array area, where internal details of the chip design are generally available, details of cell-library designs are not usually published, being much more sensitive to commercial security. Skeleton outline dimensions of polycells and full operating characteristics are usually available, so that the OEM may partition his custom design requirements into macros corresponding to the available polycells, similar to the procedure which he would follow when employing off-the-shelf SSI and MSI packages. Indeed many standard cell suppliers list the cells available in their library as equivalents of TTL 7400 or CMOS 4000 series packages, for example as shown in Table 5.8. Texas Instruments with their CMOS 74SC standard cell library go even further by using the same 74XX number as the functionally equivalent TTL package [10]. Such information in addition to the skeleton outlines therefore is the principal technical data which is commercially available for polycells.

Taking the standard cell concept to a higher level than polycells, AMI, Intel, and NCR have announced the availability of microprocessor cores as standard "cells." Other vendors are known to have similar "megacells" available or soon to be available, including PLA and memory blocks [2,9,17]. However, a danger which follows with larger-size standard parts is that they are costly to design and update and may become too specialized for widespread use, being applicable to particular requirements such as von Neumann architectures, for example. Additionally, they should be alterable in total capacity in order to meet differing custom requirements, and hence they require very intensive vendor CAD resources to alter and assemble to meet specific customer requirements. Therefore, they may be much more applicable to the vendor's own in-house design requirements, than for the wider general-purpose OEM custom market.

The availability of analog cells in the vendors' repertoire has been mentioned in technical publications, but commercial products are, as far as is known, still awaited outside the masterslice array field (see Chap. 4, Sec. 4.2.3).

TABLE 5.8 AMI Specification Sheet for CMOS Polycells Available in Their Cell Library

CMOS Standard Cells Performance Data

Combinational and I/O Elements

CELL NAME	DESCRIPTION	CMOS EQUIV	TTL EQUIV	2-IN GATE EQUIV	AREA IN$^2 \times 10^{-6}$	SPEED (TYP) nsec	POWER (TYP) @1MHz, µW
INV1	Inverter	4069	74LS04	.5	10.7	15	66
INV2	Inverting Driver	4049	74LS06	1.5	32.1	23	495
D1	Non-Inverting Driver	4050	74LS07	1.0	28.6	18	400
PD1	Non-Inverting Pad Driver W/Pad	4050	74LSO7	—	201.2	29	1850
PD1A	Non-Inverting Pad Driver W/Pad	4050	74LS07	—	233.3	29	1850
TD1	Non-Inverting 3-State Driver	40097	74LS125	2.5	42.9	26	390
TPD1	Non-Inverting 3-State Driver	40097	74LS125	2.7	227.0	41	1650
TPD1A	Non-Inverting 3-State Drive W/Pad	40097	74LS125	—	266.7	41	1650
ST1	Schmitt-Trigger CMOS Level W/Pad	40106	74LS14	—	139.3	24	150
INP1	Input Pad W/Protection Device and Inverter	4069	74LS04	—	113.5	10	150
TG2	Transfer Gate	—	—	—	14.3	32	57
ND21	2-Input NAND	4011	74LS00	1.0	14.3	19	70
ND31	3-Input NAND	4023	74LS10	1.5	25.0	19	48
ND41	4-Input NAND	4012	74LS20	2.0	21.4	31	39
ND51	5-Input NAND	4068	74LS30	2.5	28.6	31	40
NR21	2-Input NOR	4001	74LS02	1.0	14.3	23	99
NR31	3-Input NOR	4025	74LS27	1.5	17.9	28	69
NR41	4-Input NOR	4030	74LS86	2.5	28.6	32	33
ANR51	3 NAND-2 NOR Combo	4073 / 4025	74LS15 / 74LS27	2.5	25.0	34	92
OND41	2 OR + 2 NAND Combo	4071 / 4023	74LS10 / 74L32	2.0	21.4	26	47
OND42	2-2 OR – NAND	4071 / 4011	74LS00 / 74LS32	2.0	21.4	26.	72
BUS1	Bus Pad W/3-State Non-Inverting Driver + Input Inv	4097 + / 4069	74LS125 / 74LS04	—	263.1 / 288.9	13In 51Out	1650
BUS2	Bus Pad W/3-State Non-Inverting Driver + Cell Aspect Ratio	4097 / 4069	74LS125 / 74LS04	—	288.9	13In 51Out	1650
JKL2	JK Logic for D-FF	4027	74LS76	2.5	28.6	32	46

Storage Elements

CELL NAME	DESCRIPTION	CMOS EQUIV	TTL EQUIV	2-IN GATE EQUIV	AREA	SPEED	POWER
DR1	D-FF With Reset (Master-Slave)	4013	74LS74	6.0	71.4	15	486
DRS1	D-FF With Reset and Set	4013	74LS74	7.0	85.7	15	590
LSR1	D-Latch With Reset and Set	4042	74LS75	4.0	50.0	15	360
DR2	D-FF With Reset and Clock	4013	74LS74	6.0	75.0	15	460
DR3	D-FF With Reset and Clock	40174	74LS174	6.0	75.0	15	560
DRS2	D-FF With Reset, Set Clock	4013	74LS74	7.0	85.7	15	510
DFF1	D-FF With Clock	4013	74LS74	5.0	60.7	15	620

Source: Details courtesy American Microsystems Inc., Santa Clara, CA.

Second-sourcing for cell-library products currently is very problematic, because in general cell libraries are unique to particular vendors. Second-source agreements between vendors such as indicated in Chapter 4, Table 4.9 for gate-array products may evolve, but the present state of affairs is that an OEM designer (customer) is locked on to one vendor should he follow the cell-library route for the realization of his requirements. If he is a very large user, then the purchase of a complete cell library for his in-house use can be considered, which then allows him choice of mask and silicon fabrication sources. Alternatively, the use of an independent design house, not tied to a particular silicon product vendor, may offer safeguards. It is significant that there have been instances of industrial organizations outside the normal microelectronic companies who have undertaken their own in-house cell library designs in order to ensure long-term availability and independence from critical vendor situations [71]; this is expensive in terms of learning time and resources, but may be appropriate for large organizations, or even necessary where extreme end-product security or lifetime is involved. Reference to vendor literature or updated reviews such as Multi-Source N1* are necessary to establish detailed availability and interface procedures between potential customers and vendors [72].

5.5.2 Silicon Foundries

The availability of silicon foundries for the fabrication of custom-specific wafers is growing, although many of the smaller companies will perforce be behind the state of the art of the larger multinationals. This is no great disadvantage for custom-specific applications, since as has previously been mentioned, a proven and stable process is essential for good first-time results. Details of current silicon foundry availability may be found published [73]; companies will accept masks as the customer data, although some accept data base magnetic tapes of agreed format, including CIF, Calma, Applicon, and David Mann. However, it may not always be clear whether the vendor himself undertakes the magnetic tape to working plates stages, or whether he in turn may subcontract to other suppliers. The quality control which arises in the interface between design data on PG tape → mask making → specific fabrication line, with the particular idiosyncrasies of the latter, may be found discussed by Cohen and Tyree [74]; among the problems are error protection on the magnetic tape data, compensation ("bloating/shrinking") of feature sizes on working masks from the nominal dimensions to allow for fabrication line idiosyncrasies, and even difficulties in absolute measurement of a micron.

5.5.3 CAD Hardware

The availability of CAD resources for the OEM is growing fast. We have previously noted the availability of front-end data capture hardware, such as illustrated in Chap. 4, Figure 4.27. Other suppliers are also in this area. Stand-alone workstations for custom design are

---

*Multi-Source is the trademark of R. C. Anderson, Inc., Los Altos, CA.

also widely publicized, although the software to run on them for the dedication of particular cell-library or masterslice products may not be readily available. This remains one of the presently outstanding gray areas, due to the lack of standards between all sides and aspects involved.

Example sources of currently available IC design (ICD) workstations independent of any specific IC vendor include those listed in Table 5.9. For additional details see further publications [75–81]. It may be noted that there is an ongoing debate whether workstation hardware shall be dedicated hardware for the purpose of IC design, or general-purpose hardware adapted for the purpose. Both sides of this argument will be seen to be represented in Table 5.9, the arguments for and against being as follows:

For dedicated hardware: faster response time, higher performance capability for the particular job, flexibility to optimize hardware/ software and incorporate new ideas.

Against dedicated hardware: cost greater than general-purpose proprietary hardware; not so likely to be at the leading edge of hardware technology, possibly not as flexible.

For further discussions on this controversy, see Werner and elsewhere [75,80].

The Silvar-Lisco workstation has previously been illustrated (Chap. 4. Fig. 4.28). Figure 5.17 below shows the Daisy Systems Corporation LOGICAN hardware, which can be supplied as a single stand-alone workstation as shown, or paired for two-user operation on a common multibus. Its hardware features encompass:

16 Bit 10 MHz microprocessor with 0.75 to 1.5 megabytes main memory; high-performance intelligent graphics for rapid response, 17 or 19 in. graphics display.

10.5 or 42.7 Megabyte Winchester disk for on-line storage.

1.6 Megabytes unformatted or 1.0 Megabytes formatted floppy disk drive.

RS232 and/or RS422 communications interface with selectable baud rates or Ethernet.

The data base for the system supports a heirarchical structure linking behavioral simulation, logic simulation, circuit simulation, testability verification, placement and routing, and connectivity verification, with appropriate software for each activity.

TABLE 5.9 Example Third-Party Commercial Availability of Graphics-Intensive Single or Multiple-User CAD Systems for Custom-Specific IC Design Activities

Vendor	Dedicated hardware	General-purpose hardware	Interconnection flexibility
Avera Corporation[a]	√ (8086-based workstation)	—	Single-user
Cadtec Corporation	—	√ (DEC VAX; 68000 based graphics)	Up to 10 graphics stations
CAE Systems, Inc.	—	√ (Apollo)	Apollo local network
Daisy Systems Corporation	√ (8086-based workstation)	—	Single-user or pair; also Ethernet interface
Mentor Graphics Corporation	—	√ (Apollo)	Apollo local network
Metheus Corporation	√ (8086-based workstation)	—	Single-user; also Ethernet interface
Silvar-Lisco	—	√ (Apollo)	Apollo local network
Valid logic Systems, Inc.	√ (8086 or 86000 based workstation)	√ (VAX and IBM)	Single-user or cluster
Via Systems, Inc.[a]	√ (LSI 11/23)	—	Single-user

Note: not necessarily complete; subject to continuous evolution.

[a]Includes front-end data capture facilities.

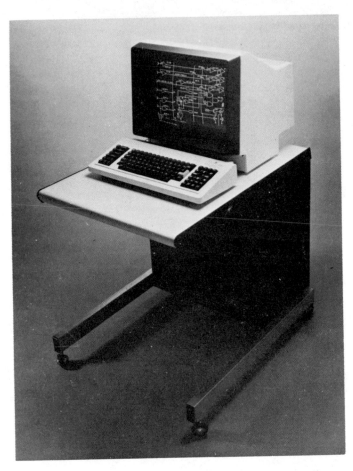

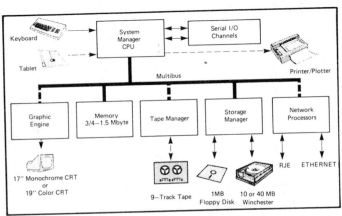

FIGURE 5.17   The Daisy LOGICIAN stand-alone workstation. (Details courtesy Daisy Systems, Inc., Sunnyvale, CA.)

These and several other workstations such as the Valid SCALD-system are powerful general-purpose LSI and VLSI design tools, which may be used by semiconductor manufacturers and other large companies for in-house custom design purposes, down to device level if required [82,83]. The SCALDsystem architecture is indicated in Figure 5.18. They are not, therefore, specifically aimed at OEM designers who may only wish to undertake the commitment design of proprietary master-slice or cell-library products for their own original equipment. However, the design and simulation tools available on such general-purpose workstations are appropriate for semicustom dedication, as may have been noted in the data contained within Figure 4.28. Similarly the Daisy LOGICAN has a development for masterslice arrays, termed the GATEMASTER, which allows automatic and/or interactive placement and routing, under the control of the central data base which maintains continuous design-rule checking of the dedication as it proceeds. The specification of the particular masterslice array being used is encoded from the vendor's detailed data, to form a specific array data base file [84].

Hence, for the OEM designer to undertake the commitment procedure for either masterslice or cell-library products, it is arguable whether full workstation capability is necessary. Certainly simulation down to device level such as provided by SPICE software (to be discussed later) is irrelevant, since the OEM is dealing with precharacterized cells or macros, and therefore the lowest level of simulation that may be required is functional-level simulation.

A step towards low-cost CAD for the OEM wishing to use standard cell semicustom ICs is provided by AMI in their SCEPTRE (Standard Cell Placement and Routing Environment) design system [85]. SCEPTRE is a software suite written in Pascal for use by the OEM, which provides placement of cells from an AMI standard cell-library, routing, and design validation. The software is designed to run on a standard configuration Sirius (Victor) microcomputer, with single-sided double-density floppy disks and 256 Kbytes of working memory. This machine in its standard form already has excellent graphics and a user-configurable keyboard.

The design operations are all menu-driven using special function keys, which are programmed to reflect the currently required options. Cells may be positioned and placed in any orientation consistant with the floor plan layout, and polysilicon and metal interconnect manually added and deleted at will. All interconnections are subject to design-rule checking to ensure that they are valid. Individual sections of the design being built up may be selected, copied, rotated, and mirrored at will.

When placement and routing is complete, 6-state logic simulation functionally checks the design, highlighting any errors or omissions. Test program vectors may be extracted, and plots of the whole or

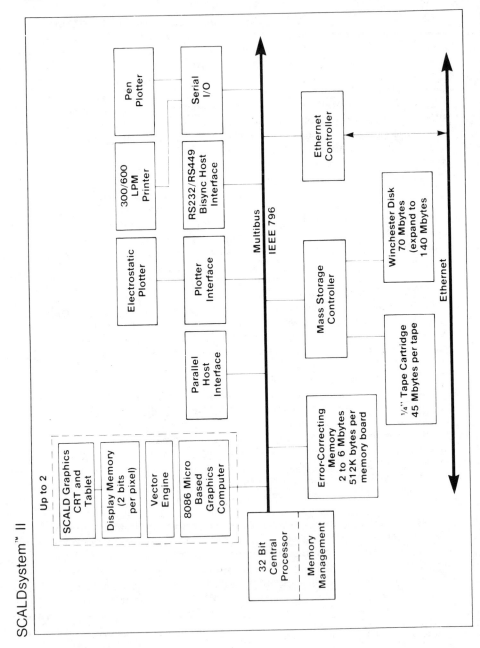

SCALDsystem™ II

selected areas of the chip may be made. On final completion, the data base disk is sent to the vendor, where the outline representations of the cells used by the OEM are replaced by the true silicon structures, and final check and fabrication is initiated.

This example, however, highlights the problem of the OEM designer when he wishes to become involved in the design cycle of a semicustom product. The outline choices available are shown in Table 5.10, ranging from full capability to do his own silicon assembly through to restrictive ties to a particular vendor and a particular product. Flexibility for the OEM designer to second-source his designs freely will be seen to be difficult, unless he is prepared to invest in what is essentially a full in-house design capability.

5.5.4 CAD Software

The range of CAD software packages written for IC design purposes is extensive, but in its own way emphasizes the lack of standardization across the whole CAD spectrum. To assemble a collection of software packages to form a comprehensive working system is difficult because it usually means modifications of interface software in order to employ a common database. Packages currently available include the following.

At simulation level, one may find a hierarchy of capability ranging through:

High-level system behavioral level

Register-transfer (data path) level

Functional-level (macro and higher) simulation

Circuit (primitive and macro) simulation

Device modelling at physics level

Process modelling

The lower of these classes is largely the province of the physicist and IC manufacturer, and is unlikely to impinge on the semicustom IC designer. The next level of simulation, however, includes ASTAP, DIANA, HILO, IMAG, MSINC, SPICE, and others to cover both logic

---

FIGURE 5.18   The SCALDsystem stand-alone workstation architecture. (Details courtesy Valid Systems, Inc., Sunnyvale, CA.)

TABLE 5.10 The Outline Decision Tree for OEM Participation in the Design of Custom-Specific Microelectronics Using CAD

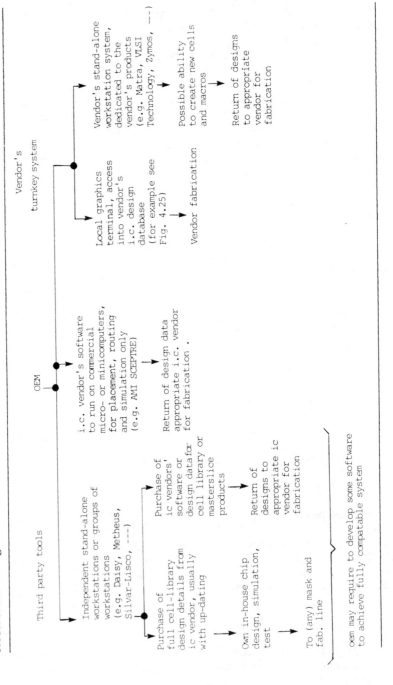

and timing simulation. Possibly the most commonly employed is SPICE, particularly the current 2G.5 version which overcomes certain short-comings present in earlier versions [2,5]. SPICE and similar programs can typically handle up to a hundred or more active devices per circuit simulation, providing ac, dc, and transient analyses. Extensions of SPICE, such as SPLICE and I-GSPICE may also be found.

At a higher functional level, suites include: DECSIM, DIGSIM, GAELIC, LOGCAP, SABLE, TEGAS, and others. Not all are commercially available, DIGSIM, LOGCAP, and TEGAS representing possibly the most commonly encountered simulators not tied to specific company use.

At the top of this heirarchical simulation tree much work is still at a research and development level, and therefore not yet widely applied. Developments such as SABLE of Stanford University are in this area, which will become much more important as VLSI utilization expands. The recently announced HELIX heirarchical simulation system of Silvar-Lisco is a significant development [86].

Layout software tends to fall into the categories of (1) in-house software of vendor's turnkey systems (see Table 5.10), (2) software written by manufacturers of independent stand-alone workstations for their product, and (3) software written by research and development groups within universities and elsewhere, which may subsequently be adopted by (1) or (2) above. Division of software into that for full-custom handcrafted design at the silicon geometry level, and placement and routing of cell-library and other structured layouts, may be made. A further category, symbolic layout, may also be noted, which involves stick diagrams or other characters, with compaction into a structured floor plan for full custom design. We will mention this further in Sec. 5.7.

In general all layout software suites are proprietary programs used in the CAD products of the various companies, for example, the SCALD graphics software of Valid Logic Systems, Inc., which itself is based upon the Berkeley University CAESAR layout program. However, a powerful recent entrant into the field is HILO-2, marketed by GenRad, which provides a complete heirarchical top-down approach, verifying each step of the design down to the gate level. Powerful new software packages such as this will become increasingly required and available.

Test generation software again is largely proprietary, for example DECSIM of Digital Equipment Corporation, but there are notable exceptions which are used by many manufacturers. The include FAIRTEST, FLASH, LOGCAP, LOGOS, SALOGS, TEGAS, and others. Possibly TEGAS is the most widely known at present.

Further details of present software packages may be found in commercial literature and elsewhere [2,5,80,81,87,88]. However, there will be increasing pressure for standardization across this whole range of software, so that transfer of data from one analysis or synthesis tool to another may be freely made. The present position is that there is no common data-exchange format, and information from, for example different front-end capture systems, are in general all dissimilar to each other.

One development towards this objective is the Texas Instruments proposal of TIDAL (Transportable Integrated Design Automation Language) as an industry standard [80], but this merely provides a means for coupling otherwise incompatible data formats, and does not truly address the more difficult problem of attempting to establish a common standard for all design and documentation data. Commercial pressures, unfortunately, may work against this objective, since it would provide the OEM with a greater freedom to shop around, not being confined so rigidly to one CAD/CAE vendor.

5.6 PACKAGING AND TESTING

Packaging of semicustom circuits has been covered in Sec. 4.9 of the previous chapter in connection with masterslice arrays. There is little further we need add here, since cell-library products have similar requirements to masterslice arrays. The only possible distinction is that masterslice arrays have a fixed complement of I/Os for each masterslice product, whereas the cell-library semicustom chip may include as many or as few I/Os as required. Therefore, there is a greater freedom in the latter to accommodate circuits which require a large number of I/Os, and hence the more recent flatpacks and pin-grid-array packages rather than dual-in-line packages may be most applicable.

Testing of masterslice arrays and cell-library circuits likewise have very similar requirements, with the latter having the potential to be larger and more complex than the former if required. Hence, any testing strategy which is applicable to cell-library designs is equally relevant for masterslice arrays, both forms of semicustom in their turn not demanding any specific testing requirements over and above standard-product testing.

A considerable range of test equipment is commercially available, ranging from simple manually operated probe-and-measure systems to fully automatic testers controlled by mainframe computers. Market leaders in sophistication include Fairchild, GenRad, Megatest, Teradine, and others. Figure 5.19 illustrates a typical large testing facility appropriate for a volume production line or commercial test house.

At the other end of the scale, wafer probes for use before packaging are widely available. Figure 5.20 shows a modern version, with 6 inch wafer capability and automatic or manual 6 inch travel in both the X and Y axis.

FIGURE 5.19   The Fairchild SENTRY 21 LSI/VLSI automatic test system, with test rates of 20 MHz in standard mode, 40 MHz in multiplex mode, 16 timing generators, 32 independent edges. (Illustration courtesy Fairchild Digital Test Systems, a Schlumberger Company, San Jose, CA.)

With the growth of on-chip circuit density both in standard products and semicustom chips, the question of design for ease of testability is assuming increasing importance. It is therefore becoming increasingly necessary to consider testability aspects of new designs at each design stage. Details of easily testable design philosophies are outside our present interests, but may be found in published literature [89—92,118] and in specialist meetings such as the annual IEEE Fault Tolerant Computing Conference.

However the possible strategies for final circuit checking will finally depend upon device structure, device size, and the number of devices to be tested. General testing strategies are summarized in Table 5.11. On the assumption that semicustom design often involves relatively small production quantities, then pressures to minimize testing time

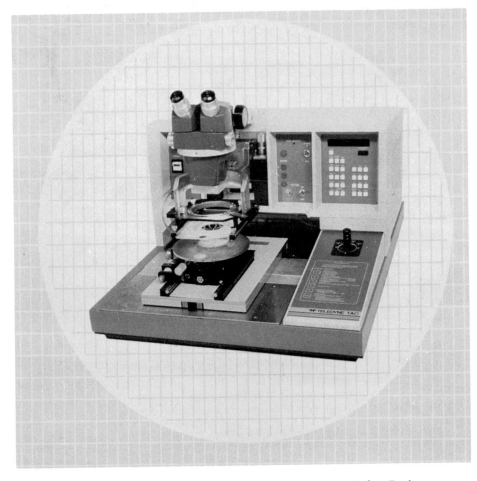

FIGURE 5.20   The Teledyne Model PR53 Automatic Wafer Prober,
microprocessor controlled, menu programmed.   (Illustration courtesy
Teledyne TAC, Woburn, MA.)

and perform less than 100% functional testing may not be so great
as in volume-product testing, and strategies to the left of Table 5.11
may be relevant.

Nevertheless, the actual machinery and implementation of testing
remains specialized, but provided the requirements of testability are
appreciated and incorporated into the custom design, then no serious
problems should be encountered in equipment availability and execution

TABLE 5.11 Heirarchy of Possible Testing Strategies

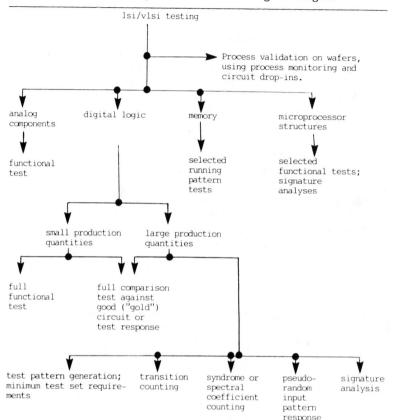

Note: reconfiguration of the circuit under test to ease testability may be present.

of semicustom product testing. For further details, see various supporting references [89−96].

5.7 FURTHER ASPECTS; FULL CUSTOM DESIGN

We have seen in the previous sections of this chapter how the original definition of the cell-library approach to semicustom design is beginning to blur. The increasing prospects of CAD software will foster developments such as:

1. The ability to modify existing cell designs to remove unwanted features, for example the $\overline{Q}$ output on latch and register elements.
2. The ability to shrink/stretch the silicon geometry of cells so as to optimize their performance for the particular application.
3. The ability to create fully parameterized new cells and macros.
4. The ability to tailor large macros (e.g., PLAs, memory) to the particular capacity required.

It should be noted that these possibilities still do not require the circuit and systems designer to be conversant with device-level design, since all the silicon design rules are contained in the data base software.

These developments may already be observed at in-house vendor level, for example in the parameterized cell compiler approach of VLSI Technology Inc. Here user-definable functional and behavioral descriptions of required cells are entered into the cell compiler CAD software to generate geometric mask data which ideally conforms to the custom-specific requirements. For example, the nMOS NOR gate compiler allows operating speeds to be chosen in five increments ranging from 2 ns minimum to 36 ns maximum, with fan-in from 1 (INVERTER) to a maximum of 8. Similar compilers customize complex macros, such as the capacity of random-access memory (RAM) and read-only memory (ROM) structures or the register length and drive capacity of shift registers, etc. Figure 5.21 illustrates how a simple noninverting buffer cell may be customized to particular on-chip requirements; additional technical information may be found published [97–103].

5.7.1 Vendor Independence

Possibilities such as those considered above still relate to a given technology and (currently) to a specific vendor's products. Of much greater significance to the OEM designer would be technology and vendor independence at the design stage, with the resultant freedom for the OEM to acquire *functionally interchangable* custom-specific circuits from several IC vendors.

Therefore, we may postulate a technology-independent and vendor-independent design data base, with which OEM designers may complete their functional designs, including all performance requirements, at the completion of which vendor-specific data base information may be called in to complete a specific vendor product. This is illustrated in Table 5.12. A great deal of development work will be involved in the establishment of such a design system, with possibly mixed enthusiasm from the independent IC vendors.

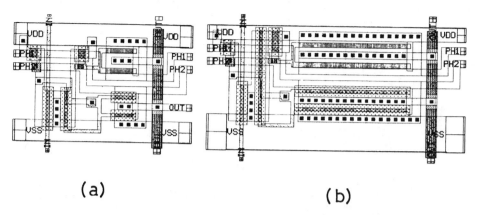

(a)   **(b)**

FIGURE 5.21   The generation of application-specific cells from standard data base cell-library specifications: (a) simple noninverting
buffer to drive 0.5 pF loading in 20 ns, (b) an application-specific version, to provide fast-rising, slow-falling response when driving 10 pF
loads. (Details courtesy VLSI Technology Inc., San Jose, CA.)

5.7.2 Full Custom Considerations

All the preceding considerations have been IC design techniques and
philosophies which involve some form of predesigned material, basically
in the form of uncommitted array or cell-library approaches. We have
not specifically considered chip design which lies outside these areas,
in what may be considered full custom.

In bipolar technology, it is generally still true that full custom is
the province of the design experts, the "toolroom" designer, within
an IC manufacturing company. Due to the generally more complex fabrication rules compared with MOS, and emphasis upon maximum available
performance, OEM designers unfamiliar with full technology details are
unlikely to become involved in bipolar chip design. However nMOS and
CMOS allows simplification; design rules based upon scaling and aspect
ratios can be sufficiently defined so that the OEM designer can, with
CAD support, undertake what may be considered full-custom activities.
However, we must treat the terms "semicustom" and "full custom" with
care, since in the area which we have just defined as full custom the
designer is still constrained by the technology design rules laid down
by the toolroom silicon designer.

TABLE 5.12 Possible Future Vendor-Independent Cell-Library Application Specific Design Procedure

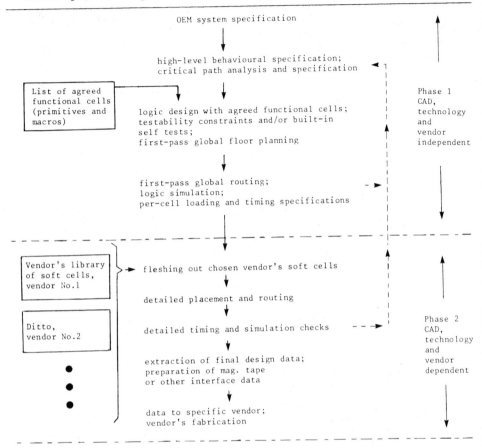

The terminology "silicon compiler" is frequently encountered in this area beyond the polycell approach, although it is extremely difficult to agree upon just what this term means in practice. In theory silicon compilation is the automatic generation of a complete silicon floor plan layout from a behavioral description of the required system, using data base information on silicon design rules and performance. To quote A. M. Peskin in the 1982 IEEE Customs Integrated Circuits Conference:

> The term silicon compiler is used to denote a program with the translational attributes of a compiler, yet whose objective is a hardware device rather than software. It is, in a sense, the ultimate CAD tool.

Thus the ideal silicon compiler would automatically span the complete range of activity:

High level functional system description

↓

Register transfer language description

↓

Boolean or equivalent functional descriptions

↓

Logic diagram or equivalent description level

↓

Test generation data

↓

Geometric floorplan layout for mask fab.

Indeed some may aruge that no intermediate descriptive levels from register transfer language to layout are necessary.

Thus, in theory, silicon compilers would determine the complete floor plan layout of an IC, with no preconceived structural constraints, directly generating the necessary fabrication mask descriptions. The arguments propounded are that such designs will be correct by construction, thus giving first-time-right capabilities; against this the utilization of silicon area will be inefficient. In practice, developments to date are much more module or silicon *assemblers*, since their capabilities lie in the on-chip assembly of well-defined but possibly variable-size structures such as PLA's and shift registers.

For further debate and discussion on this currently contentious term, introduced largely by computer scientists, see Werner [98] and other publications [5,99–107,116]. The titles of many of these references indicate the range of activities with which the term "silicon compiler" has been associated. Further comments will be found by Denyer [5] and in Mavor et al. [4]. In contrast, for a comprehensive discussion of CAD systems for IC design which does not use this term, see Daniel and Gwyn [19].

There are, however, two presently proposed methods of full custom down to device level in which the OEM designer may participate, given the appropriate CAD support; the first is the "Bristle-Block" approach originated by the California Institute of Technology (CALTECH) for data path architectures, the second being the Mead-Conway structured approach, used widely in graduate educational courses on IC design. Both are in *n*MOS technology, although extensions to CMOS have been considered.

The Bristle-Block approach does not cover the custom design of a complete chip, but only the data path sections of microprocessor type structures. Thus registers, arithmetic logic units and data-shifting networks operating on data highways can be accommodated, but not the circuitry for the microsequencer and other more random logic requirements. The latter require other realization means, for example PLA structures.

Given that the required system is a data bus-intensive type structure, the Bristle-Block approach is to assemble appropriate functional core elements horizontally along each data path, the elements being stacked in parallel across the several parallel data paths to form macros. The core elements themselves are parameterized primitives stored in outline but not detailed layout form. The design procedure gives the width of the data paths present (for example 16 bits) and the microcode conditions for each series element so as to compile the required overall operation. The CAD program then calls up the required core elements, and commences a check on the design dimensions of each. Cell stretching is then performed so that cells along each data path can directly abut, the output connections from one directly aligning with the inputs to the next, thus obviating the need for any horizontal interconnection design. This is illustrated in Figure 5.22a which shows the stretching of verical connections within the cells to achieve the horizontal alignment. Hence, a fully abutting floor plan is built up, as shown in Figure 5.22b.

Further details of this technique may be found published [4,7,101, 108]. It will be appreciated that this is a "soft cell" technique, which may be particularly powerful for data path architectures, but not for all OEM applications. Further work on data path system design is reported by IBM and Bell Labs [109].

The Mead-Conway structured approach to custom design is based upon a set of nMOS geometric design rules, which are simplifications of the various masking stages used in the production process. These design rules are "portable," that is they have been generalized so as to be applicable to any conventional nMOS production line, not being vendor-specific, and hence not involving ultimate state-of-the-art fabrication techniques which may not be portable.

The principal rationalization introduced by Mead and Conway involves normalization of the critical geometric dimensions involved in nMOS silicon layouts, in terms of a basic dimensional parameter lambda (λ). This parameter λ encompasses the critical dimensions present in any nMOS fabrication, such as line widths and spacings, mask alignment tolerance, stretch and shrink between mask and actual fabrication dimensions, overlaps between polysilicon and diffusion, and so on, and may therefore be viewed as the parameter by which all geometric details may be defined. Typically λ may be, say, 2 microns, and feature details expressed as 2λ, 4λ, etc., λ can then be given a specific value for any

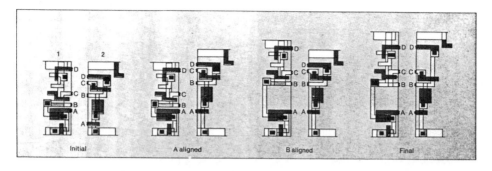

(a)

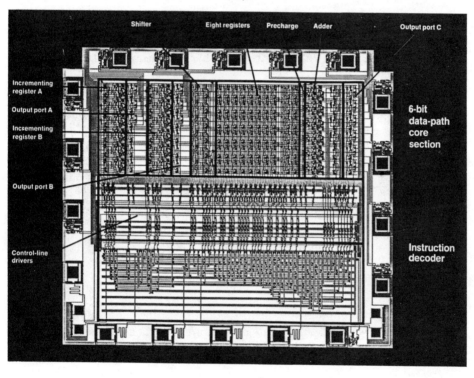

(b)

FIGURE 5.22   The Bristle-Block technique developed by CALTEC:
(a) cell abuttment by stretching, (b) floor plan layout. (Acknowledgements Trimberger, S., Automating chip layout. *IEEE Spectrum*, Copyright © 1982, IEEE.)

given production line by the particular IC manufacturer. It is, therefore, a very conservative overall parameter which can characterize a
production process, but which allows independent groups to design and
lay out nMOS circuits, the appropriate λ value being incorporated immediately prior to specific mask making. It also allows different groups
to design independently of each other, amalgamation of several different
designs on one die (multiproject chips, or MPCs) being possible—this
is the procedure widely used in design courses in educational establish-
ments, which in turn accounts for the emphasis upon the "standardized"
Mead-Conway approach in the educational field. Hence, the principal
advantages and disadvantages of the Mead-Conway approach are:

Advantages:

1. Generous design rules, so that final fabrication can be undertaken by any nMOS fabrication line.
2. Scaleability (within limits), so that with improved fabrication
   availability device and circuit sizes may be shrunk.
3. Rationalization of design rules, which is advantageous in an
   educational environment.
4. Portability of the design rules and supporting CAD software.

Disadvantages:

1. Silicon area employed is considerably higher than other nMOS
   custom designs, possibly up to 50% higher, since the common λ
   value is overgenerous in most cases.
2. Does not provide state-of-the-art circuit performance because of
   its conservative design rules.
3. Confined (at present) to nMOS technology.
4. Not used in commercial areas, largely because of loss of commercial edge due to 1 and 2 above.

It therefore represents a powerful vehicle for the practical introduction
of IC design to students, helped by the availability and interchange
of CAD software and other material. However, it cannot entirely keep
pace with developments in reduced silicon geometries, since the single
λ factor becomes increasingly inadequate as possible feature sizes shrink
towards the micron level.

The standard bible for this approach is the Mead and Conway textbook
[1], which is widely known. The stick diagrams representing the basic
nMOS logic circuits and the structured architectural design philosophy
are fully illustrated in this text. Examples of the silicon layout design
rules are shown in Figure 5.23a, but color plates are necessary to show
more explicitly the overlap of the various processing levels of a full
nMOS circuit. Figure 5.23b attempts to give a NAND gate representation
within present printing constraints.

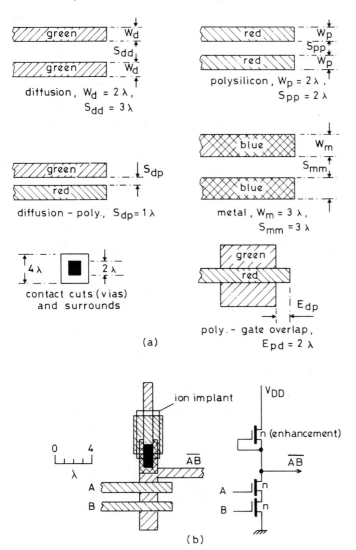

(a)

(b)

FIGURE 5.23 The Mead-Conway custom approach: (a) the geometric design rules, giving diffusion and polysilicon dimensions, spacings and overlaps in terms of λ, (b) NAND gate layout geometry.

The design tools available with this approach include the Caltech Intermediate Form (CIF) software. This is a means of describing geometric features involved in the nMOS designs, and serves as a standard machine-readable representation of the geometries with which video display units (VDUs), plotters, and pattern-generation magnetic tapes may be controlled. CIF therefore acts as the intermediate standard form in the design phase before translation into whatever variety of formats or display may be required. Thus it will be appreciated that the Mead-Conway approach is full-custom design down at geometric device level, not employing any predesigned cells or macros, but relying upon the few but sufficient design-rule checks which are possible with the simple structured assembly of ratioed nMOS circuits. Its use in multiproject chips may be found discussed in several publications [110,111].

Extension of portable design rules to CMOS has received consideration [112,113]. Undoubtedly we will see this development spreading in the educational field, but like the present nMOS situation it will inevitably involve conservative rules and regulations in order to achieve portability.

Finally, a brief mention of special-purpose LSI/VLSI chips, custom designed to realize specific signal processing algorithms or similar duties. Clearly, if the requirements have some iterative functional structure, as may be present in pattern matching, recursive digital filtering, discrete Fourier transforms, linear system solving, and the like, then a one-to-one mapping from the computational structure to an array of appropriate functional cells readily follows. The possibilities of the structured design of special-purpose computational chips, involving the direct realization of the mathematical algorithm in functional cell and chip layout, may be found discussed by Foster and Kung [114] and by McCabe et al. [115]. A good introduction to systolic arrays may be found in the latter.

5.8 SUMMARY

In the preceding sections, we have attempted to review the current state of the art in cell-library-based custom-specific design. It has been seen that not such a diverse range of vendor products for OEM use is currently available compared with the masterslice alternatives, but the inherent greater flexibility and performance of the cell-library approach promises greater future adoption and use, particularly with the increasing power of CAD resources.

At the conclusion of Chapter 4 we listed five problems of current concern in the masterslice-array area; these five points are equally if not more significant in the cell-library area, added to which must be the present complete lack of portability of commercial software and standardization of data formats. This is the most serious problem currently facing the original equipment manufacturer who wishes to become involved in the design cycle of his own custom-specific circuits, one which it may take considerable time to resolve.

However, without any doubt, cell-library and its many and varied derivatives will emerge as the most significant design philosophy for LSI and VLSI in the future. Its use in the realization of special-purpose signal-processing chips seems particularly advantageous, although the design of sophisticated signal-processing cells may require toolroom design. The following chapter will attempt a final overview of all the present and the future prospects.

REFERENCES

1. Mead, C., and Conway, L. *Introduction to VLSI Systems.* Addison-Wesley, Reading, MA, 1980.
2. Muroga, S. *VLSI System Design.* Wiley-Interscience, New York, 1982.
3. Howes, M. J., and Morgan, D. V. (Ed.). *Large Scale Integration.* John Wiley, London, 1981.
4. Mavor, J., Jack, M. A., and Dyner, P. B. *Introduction to MOS LSI Design.* Addison-Wesley, Chichester, UK, 1983.
5. Hicks, P. J. (Ed.). *Semi-Custom IC Design and VLSI.* IEE Peter Peregrinus, London, 1983.
6. Ramsey, F. Autolayout of ULA gate arrays on an engineer's work station. *Proc. IEEE Int. Symp. on Circuits and Systems,* paper S.19/3, 1983.
7. Brown, M. A. CMOS cell arrays—an alternative to gate arrays. *Proc. IEEE Custom Integrated Circuits Conf.,* 74–76, May 1983.
8. Pritchard, R. L. Gate arrays or standard cells. *J. Semi-Custom IC's,* 1(3):5–8, 1984.
9. Technology Focus. The standard cell design option. *Electronic Engineering,* 55(678):135–138, 1983.
10. Rappaport, A. EDN semi-custom ic directory. *EDN,* 28(4):78–218, 1983.
11. Rappaport, A. EDN semi-custom ic directory. *EDN,* 29(4):91–242, 1984.
12. Twaddell, W. Standard-cell design inroads spur specialized gate growth. *EDN,* 28(22):49–58, 1983.
13. ICE Report. *Status 1984: A Report on the Integrated Circuit Industry.* Integrated Circuit Engineering Corporation, Scottsdale, AZ, 1984.

14. Report. *Standard Cell Custom IC Report*. HTE Management Resources, Sunnyvale, CA, 1982.

15. Szirom, S. Picking the right custom ic solution. *Defense Electronics*, 96–101, September 1982.

16. AMI. Standard cell custom design. Technical brochure 05-1-82 10M, American Microsystems Inc., Santa Clara, CA.

17. Loesch, W. Custom ic's from standard cells: a design approach. *Computer Design*, 227–234, May 1982.

18. Walsh, M. The alterable microcomputer adapts to complexity. *Electronic Engineering*, 55(678):141–144, 1983.

19. Daniel, M. E., and Gwyn, C. W. CAD systems for IC design. *IEEE Trans.*, CAD.1:2–12, 1982.

20. Feller, A., and Agostino, M. D. Computer-aided mask artwork generation for IC arrays. *Digest IEEE Computer Group Conf.*, 23–26, 1968.

21. Feller, A. Automatic layout of low-cost quick turn-around random logic custom lsi circuits. *Proc. 13th IEEE Design Automation Conf.*, 79–85, 1976.

22. Persky, G., Deutsch, D. H., and Schweikert, D. G. LTX—a system for the directed automatic design of LSI circuits. *Proc. 13th IEEE Design Automation Conf.*, 399–407, 1976.

23. Jennings, P. I., Hurst, S. L., and McDonald, A. A highly routable gate array and its automated customization. *IEEE Trans.*, CAD.3:27–40, 1984.

24. Hanan, M., and Kurtzberg, J. M. Placement techniques. In *Design and Automation of Digital Systems, Theory and Techniques*. Ed. M. A. Breur. Prentice-Hall, Englewood Cliffs, NJ, 1972.

25. Kozawa, T. et al. Automatic placement algorithms for high packing density vlsi. *Proc. IEEE 20th Design Automation Conf.*, 175–181, 1983.

26. Kodres, U. R. Formulation and solution of circuit card design problems through the use of graph methods. In *Advances in Electronic Circuit Packaging*. Ed. Walker, G. A., Plenum Press, New York, 1962.

27. Pomentale, T. An algorithm for minimizing backboard wiring functions. *Communciations ACM*, 8:699–702, 1965.

28. Kernighan, B. S., and Lin, W. An efficient heuristic procedure for partitioning graphs. *Bell System Technical Journal*, 49:291–307, 1970.

29. Schweikert, D. G. A 2-dimensional placement algorithm for the layout of electrical circuits. *Proc. IEEE 13th Design Automation Conf.*, 408–415, 1976.

30. Hanan, M., Wolff, P. K., and Anguli, B. J. A study of placement techniques. *J. Design Automation Fault Tolerant Computing*, 1:28–61, 1976.

31. Khokhani, K. H., and Patel, A. M. The chip layout problem: a placement procedure for LSI. *Proc. IEEE 14th Design Automation Conf.*, 291–297, 1977.
32. Breur, M. A. Min-cut placement. *J. Design Automation Fault Tolerant Computing*, 1:343–362, 1977.
33. Schmidt, D. C., and Druffel, L. E. An iterative algorithm for placement and assignment of integrated circuits. *Proc. IEEE 12th Design Automation Conf.*, 361–368, 1975.
34. Lauther, U. A min-cut placement algorithm for general cell assemblies. *J. Design Automation Fault Tolerant Computing*, 4:21–34, 1980.
35. Steinberg, L. The backboard wiring problem: a placement algorithm. *SIAM Review*, 3:37–50, 1961.
36. Rutman, R. A. An algorithm for placement of interconnected elements based upon minimum wire lengths. *Proc. AFIPS Conf. SJCC*, 477–491, 1964.
37. Wilson, D. C., and Smith, R. J. An experimental comparison of force directed placement techniques. *Proc. IEEE 11th Design Automation Workshop*, 194–199, 1974.
38. Fisk, C. J., Caskey, D. C., and West, L. L. Automated circuit card etching layout. *Proc. IEEE*, 55:1971–1982, 1967.
39. Shiraishi, H., and Hirose, F. Efficient placement and routing techniques for masterslice lsi. *Proc. IEEE 17th Design Automation Conf.*, 458–463, 1980.
40. Persky, G. PRO—an automatic string placement program for polycell layout. *Proc. IEEE 17th Design Automation Conf.*, 417–424, 1976.
41. Kawanishi, H., Yoshizawa, H., and Kani, K. A heuristic procedure for ordering MOS arrays. *Electronics Communications Japan*, 59A:51–59, 1976.
42. Beke, H., and Sansen, W. CALMOS—a portable software system for the automatic and interactive layout of MOS/LSI. *Proc. IEEE 16th Design Automation Conf.*, 102–108, 1979.
43. Kamikawa, R., Osawa, A., Yasunda, I., and Cliba, T. Placement and routing program for masterslice lsi. *Proc. IEEE 16th Design Automation Conf.*, 245–250, 1976.
44. Berge, C. *The Theory of Graphs and its Application.* John Wiley, New York, 1962.
45. Dantzig, G. *Linear Programming and Extensions.* Princeton University Press, Princeton, NJ, 1963.
46. Harray, F. *Graph Theory.* Addison-Wesley, Reading, MA, 1969.
47. Chyan, D-Y., and Breur, M. A. A placement algorithm for array processors. *Proc. IEEE 20th Design Automation Conf.*, 182–188, 1983.
48. Supowit, K. J., and Slutz, E. A. Placement algorithms for vlsi. *Proc. IEEE 20th Design Automation Conf.*, 164–170, 1983.

49. Weisel, M., and Mlynski, D. A. An efficient channel model for building block lsi. *Proc. IEEE ISCAS*, 118–121, 1981.
50. Rothermel, H-J., and Mlynski, D. A. Routing method for vlsi design using irregular cells. *Proc. IEEE 20th Design Automation Conf.*, 257–262, 1983.
51. Horng, C., and Lie, M. An automatic/interactive layout planning system for arbitrarily sized rectangular building blocks. *Proc. IEEE 18th Design Automation Conf.*, 293–300, 1981.
52. Preas, B. T., and vanCleemput, W. M. Placement algorithms for arbitrary shaped blocks. *Proc. IEEE 16th Design Automation Conf.*, 474–480, 1979.
53. Tarai, H. Automatic placement and routing program for logic vlsi design based upon hierarchical layout method. *Proc. IEEE Int. Conf. Circuits and Computers*, 415–418, 1983.
54. Akers, S. B. Routing. In *Design and Automation of Digital Systems, Theory and Techniques*. Ed. M. A. Breur. Prentice-Hall, Englewood Cliffs, NJ, 1972.
55. Hightower, D. W. The interconnection problem: a tutorial. *Computer*, 7:18–32, 1974.
56. Lee, C. Y. An algorithm for path connection and its application. *IEEE Trans.* EC.10:346–365, 1961.
57. Akers, S. B. A modification of Lee's path connection algorithm. *IEEE Trans.*, EC.16:97–98, 1967.
58. Hoel, J. H. Some variations of Lee's algorithm. *IEEE Trans.* C.25:19–24, 1976.
59. Hightower, D. W. A solution to line-routing problems on the continuous plane. *Proc. IEEE 6th Design Automation Conf.*, 1–24, 1969.
60. Heyns, W., Sansen, W., and Beke, H. A line expansion algorithm for the general routing problem, with a guaranteed solution. *Proc. IEEE 17th Design Automation Conf.*, 243–249, 1980.
61. Hashimoto, A., and Stevens, J. Wire routing by optimizing channel assignment within large apertures. *Proc. IEEE 8th Design Automation Conf.*, 155–169, 1971.
62. Jennings, A. A topology for semicustom array-structured lsi devices and their automatic customization. *Proc. IEEE 6th Design Automation Conf.*, 675–681, 1983.
63. Kernigham, S., Schweikert, D., and Persky, G. An optimum channel-routing algorithm for polycell layouts of integrated circuits. *Proc. IEEE 10th Design Automation Conf.*, 50–59, 1973.
64. Deutsch, D. N. A dog-leg channel router. *Proc. IEEE 13th Design Automation Conf.*, 425–433, 1976.
65. Yoshimura, T., and Kuh, E. S. Efficient algorithm for channel routing. *IEEE Trans.*, CAD.1:25–35, 1982.
66. Rivest, R. L., and Fiduccia, C. M. A "greedy" channel router. *Proc. IEEE 19th Design Automation Conf.*, 418–424, 1982.

67. Burstein, M., and Pelavia, R. Hierarchical channel router. *Proc. IEEE 20th Design Automation Conf.*, 591–597, 1983.

68. Leong, H. W., and Liu, C. L. A new channel routing problem. *Proc. IEEE 20th Design Automation Conf.*, 584–590, 1983.

69. Deutsch, D. N., and Glick, P. An over-the-cell router. *Proc. IEEE 17th Design Automation Conf.*, 32–39, 1980.

70. Krohn, H. E. An over-the-cell gate array channel router. *Proc. IEEE 20th Design Automation Conf.*, 665–670, 1983.

71. Gunn, I. British Aerospace own 3 µm CMOS process cell array and integrated design system. *Proc. 3rd Int. Conf. Semi-Custom ICs*, paper2/1, London, 1983.

72. Anderson, R. C. *Multi-Source N1; Integrated Circuit Design and Procurement System.* R. C. Anderson, Los Altos, CA, 1982, (with updates).

73. Review. A survey of silicon foundries. *VLSI Design*, 3:42–49, July 1982.

74. Cohen, H., and Tyree, V. Quality control from the silicon broker's perspective. *VLSI Design*, 3:24–30, July 1982.

75. Werner, J. Sorting out the CAE workstations. *VLSI Design*, 4:46–55, March 1983.

76. Robson, G. Benchmarking the workstations. *VLSI Design*, 4:58–61, March 1983.

77. Hiller, W. E. The computer-aided engineering workstation. *IEE Electronics Power*, 29:69–71, January 1983.

78. Broster, D., and South, A. The role of CAD in semi-custom ic design. *IEE Electronics Power*, 29:42–46, January 1983.

79. Harrow, P. The selection of CAD systems. *IEE Electronics Power*, 29:63–68, January 1983.

80. Rappaport, A. Computer-aided engineering tools. *EDN*, 28(19): 106–138, 1983.

81. Report. Computer aided engineering workstations for the ic industry. Strategic Inc., San Jose, CA, 1983.

82. Jenné, D. C., and Stamm, D. A. Managing VLSI complexity: a user's viewpoint. *VLSI Design*, 3:14–20, March 1982.

83. Miller, D., and Miranker, G. Terminal-based engineering system cuts logic design time tenfold. *Electronics*, 55(18):135–139, 1982.

84. King, R. Universal approach to gate array development. *Proc. 3rd Int. Conf. on Semi-Custom IC's*, paper 3/1, London, 1983.

85. Walsh, M. A low cost semi-custom workstation. *Proc. 3rd Int. Conf. on Semi-Custom IC's*, paper 3/2, London, 1983.

86. Coelho, D. R., and vanCleemput, W. M. HELIX, a tool for multi-level simulation of VLSI systems. *Proc. 3rd Int. Conf. on Semi-Custom IC's*, paper 2/5, London, 1983.

87. Werner, J., and Beresford, R. A system engineer's guide to simulators. *VLSI Design*, 5(2):27–33, 1984.

88. Musgrave, G. (Ed.). *Computer-Aided Design of Digital Electronic Circuits and Systems.* North-Holland, Amsterdam, 1979.

89. Bennetts, R. G. *Design of Testable Logic Circuits.* Addison-Wesley, Reading, MA, 1984.

90. Williams, T. W., and Parker, K. P. Design for testability: a survey. *Proc. IEEE,* 71:98−112, 1983.

91. Twaddell, W. Semi-custom logic suppliers differ how to best deal with testability. *EDN,* 27(23):69−73, 1982.

92. Muehldorf, E. I. High-speed integrated circuit characterization and test strategy. *Solid State Technology,* 93−97, September 1980.

93. Singer, P. H. VLSI test systems: facing the challenge of tomorrow. *Semiconductor International,* 7(1):82−88, 1984.

94. Turino, J. Trends in lsi/vlsi testing. *Semiconductor International,* 7(1):70−75, 1984.

95. Walker, R., and King, D. Testing logic arrays. *Microelectronics J.,* 14(3):31−39, 1983.

96. Grassl, G., and Pfleiderer, H. -J. A function independent self-test for large programmable logic arrays. *Integration, VLSI J.,* 1(1):71−80, 1983.

97. Nance, S., Starr, C., Dunn, R., and Kliment, M. Cell layout compilers simplify custom ic design. *EDN,* 28(19):147−158, 1983.

98. Werner, J. The silicon compiler—panacea, wishful thinking, or old hat. *VLSI Design,* 3:46−52, September 1982.

99. Ayres, R. Silicon compilation—a hierarchical use of PLA's. *Proc. IEEE 16th Design Automation Conference,* 314−326, 1979.

100. Gray, J. P. Introduction to silicon compilation. *Proc. IEEE 16th Design Automation Conf.,* 305−306, 1979.

101. Johannsen, D. L. Bristle blocks—a silicon compiler. *Proc. IEEE 16th Design Automation Conf.,* 310−313, 1979.

102. Gray, J. P., Buchanan, I., and Roberton, P. S. Designing gate arrays using a silicon compiler. *Proc. IEEE 19th Design Automation Conf.,* 377−383, 1982.

103. Luhakay, J., and Kubitz, W. J. Layout synthesis system for nMOS gate cells. *Proc. IEEE 19th Design Automation Conf.,* 307−314, 1982.

104. Peskin, A. M. Towards a silicon compiler. *Proc. IEEE Custom Integrated Circuits Conf.,* 125−128, 1982.

105. Reiss, S., and Savage, J. SLAP, a silicon layout program. *Proc. IEEE Int. Conf. Circuits and Computers,* 281−284, 1982.

106. Deyner, P. B., Renshaw, D., and Begmann, N. A silicon compiler for VLSI signal processors. *Proc. European Solid-State Circuits Conf.,* (ESSCIRC 82), Brussels, 1982.

107. Fett, D., Loeffler, D., Nielsen, D., and Tobias, J. Using a silicon compiler to generate ic test structures. *VLSI Design,* 3:56−58, September 1982.

108. Trimberger, S. Automating chip layout. *IEEE Spectrum*, 19: 38—45, June 1982.
109. Tobias, J. R. LSI/VLSI building blocks. *IEEE Computer*, 14(8):83—101, 1981.
110. Conway, L., Bell, A., and Martin, E. N. A large-scale demonstration of a new way to create systems in silicon. *Lambda* (now *VLSI Design*), 1(2):10—19, 1980.
111. Lyon, R. F. Simplified design rules for VLSI layouts. *Lambda*, 2(1):54—59, 1981.
112. Griswold, T. W. Portable design rules for bulk CMOS. *VLSI Design*, 3:62—67, September 1982.
113. Lipman, J. A CMOS implementation of an introductory VLSI design course. *VLSI Design*, 2(4):56—58, 1981.
114. Foster, M. J., and Kung, H. T. The design of special-purpose chips. *VLSI Design*, 1(1):26—40, 1980.
115. McCabe, M. M., Arambepola, B., and Corry, A. G. New algorithms and architectures for VLSI. *GEC J. Science Technology*, 48(2):68—75, 1982.
116. Wallich, P. On the horizon: fast chips quickly. *IEEE Spectrum*, 21(3):28—34, 1984.
117. Robson, G. Automatic placement and routing of gate arrays. *VLSI Design*, 5(4):35—43, 1984.
118. Bennetts, R. G. Practical guidelines for designing testable custom/semicustom ICs. *VLSI Design*, 5(4):64—71, 1984.

6
Conclusions

6.1 THE INCREASING ADOPTION OF CUSTOM DESIGN

We have now completed our coverage of the present-day custom and semicustom scene. We have seen that although programmable off-the-shelf devices such as programmable logic arrays (PLAs) and programmable array logic (PALs) have a useful role in the system designer's armory, it is the masterslice gate array and the cell-library area which have the greatest significance, both now and in the foreseeable future. In this concluding section we will attempt a look into future developments in these two areas, and indicate that a shift away from gate array for production equipments may be the future tendency, this shift being fueled by increasingly available computer-aided design (CAD) resources.

However, whatever the exact balance between the available custom and semicustom techniques may be at any particular time, it is firmly established that custom-specific microelectronics will be the microelectronics growth area in the coming decade. Off-the-shelf standard products will take a reduced percentage of the market, but possibly not in terms of absolute value due to the general increase in total world consumption.

Figure 6.1 illustrates forecast figures for the rest of the present decade. The total custom-specific market is forecast to exceed $2000 million by 1989 from the 1982 figure of under $200 million. These values include all forms of custom and semicustom products.

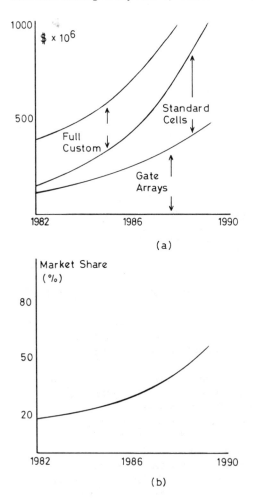

FIGURE 6.1   Forecast growth in custom-specific microelectronics:
(a) annual global sales, (b) percentage of market served by standard
cells.

Further analysis behind these figures suggests that: (1) the
masterslice gate-array market will show a 35% annual growth rate,
(2) standard cells will show a 55% annual growth rate, and (3) standard
cells will expand their share of the total custom-specific market to above
60% by 1989.

The number of gates required by the majority of custom-specific
applications will grow from the present value of less than 1000, but is

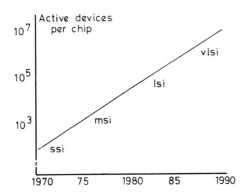

FIGURE 6.2   The increase in available devices per chip.

unlikely to exceed 2000 for the vast majority of real-life applications.
Similarly, the currently available speed from modern 3 μm CMOS
should be adequate to cater for greater than 90% of all anticipated
applications. Higher speeds may be required for sophisticated circuits, such as on-line signal processing and mainframe computing
applications.

The increase in the number of possible active elements (transistors)
per chip is shown in Figure 6.2. This expansion in absolute capability
will be absorbed by the standard off-the-shelf memory market, but is
unlikely to be required in custom-specific chips. However, this capability will spin off into the smaller size chips, resulting in improved
fabrication and yields for smaller products.

Looking specifically at masterslice gate arrays, the state-of-the-art
technology which allows greater chip area and smaller device geometry
is reflected in the graphs of Figure 6.3. Commercially available production lines may lag behind these time scales by about two years. However, as we have noted above, it is considered unlikely that the majority of masterslice gate-array applications will require this increased
absolute capability.

Standard cells will similarly show an increased capacity per chip, as
illustrated in Figure 6.4. Unlike gate arrays, there may well be a percentage of standard-cell and full custom applications which will employ
the full technological capability of the moment. Note that increasing
design ability, including improved CAD and the introduction of "soft
cells" (to be discussed later), will close the gap between standard cells
and handcrafted designs.

The breakdown by process technology of the custom-specific market
is forecast to be as follows:

	Present (1983)	Future (1990)
CMOS	35%	80%
Emitter-coupled logic (ECL)	35%	15%
Others	30%	5%

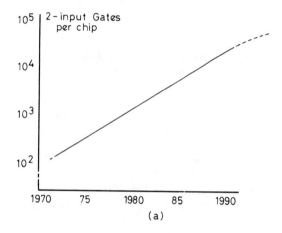

(a)

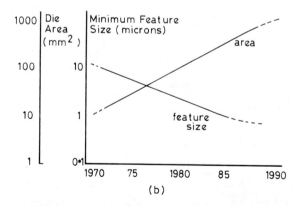

(b)

FIGURE 6.3   State-of-the-art technology increases for masterslice gate arrays: (a) number of possible gates per chip, (b) feature size and chip area availability.

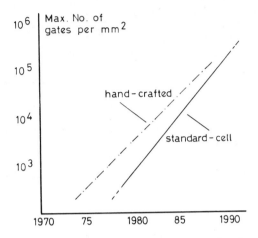

FIGURE 6.4   State-of-the-art technology increases for standard cell design; handcrafted capability included for comparison purposes.

50% of all present-day custom-specific applications have requirements which demand less than one-third of the available maximum gate-count and gate-speed availability in CMOS: this factor is likely to be unchanged in the future decade, since CMOS improvements will keep step with the gradual growth in application requirements.

   Packaging, however, for the custom-specific market may require greater emphasis and innovation than new processing technology. Trends as follows are forecast in packaging requirements:

	Present (1983)	Future (1990)
14 and 16 pin DIL	30%	15%
18 to 40 pin DIL	40%	25%
14 to 40 pin leadless ceramic	20%	20%
Others, including plastic leadless	10%	40%

   The above forecasts of market expansion and subdivision between gate arrays, cell library, and full custom are placed very much in question by the future growth and adoption of CAD workstations.   The

potential explosion in availability and capability in this area may have
the following effects which so far have been underestimated by published forecasts:

1. An increased total in custom-specific microelectronic production
   over and above that forecast to date.
2. A more rapid decline in the percentage of the total custom-
   specific market covered by masterslice gate arrays.
3. An increase in the percentage of the custom-specific market
   covered by full custom, with a blurring of the division between
   cell-library and full-custom design.

We will consider the latter point in the following section of this chapter.
However, whatever may be the exact influence of CAD workstations,
in total they will undoubtedly accelerate the use of custom-specific
microelectronics and reduce design times to well below current industry
standards. Possible design times to prototype samples may be as follows:

	Present (1983)	Future (1990)
Full custom	$\geqslant$50 weeks	2/3 weeks
Cell library	30 weeks	2/3 weeks
Gate array	20 weeks	2 weeks

The bottleneck may then lie in mask making rather than design and
fabrication, but this in turn may be relieved by direct write-on-wafer
or other developments.

Further forecast details may be found published [1-11]. However,
the future of custom-specific microelectronics in one or another form
is assured.

6.2 FUTURE DEVELOPMENTS

Within the above framework of a generally anticipated increase in microelectronics in forthcoming years let us finally make some comments
and observations on technological bounds and specific developments.

6.2.1 General Bounds

Figure 6.5 shows the general evolutionary trend of optimum scale of
integration with time. At any point in time the maximum possible chip

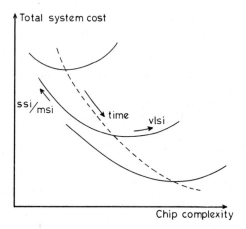

FIGURE 6.5   The optimum scale of integration for OEM use versus time.

complexity is greater than the original equipment manufacturer (OEM) can profitably use, the optimum scale of integration for minimum system cost being somewhere between low-complexity chips and the maximum currently possible. It is difficult to quantify such a general graph, but it represents a well-established economic equation, and is likely to

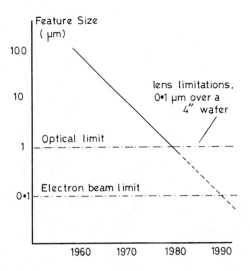

FIGURE 6.6   The limits of optical lithography. (Note: mask alignment and etching accuracy are separate limitations.)

continue in the future—it merely emphasizes that the latest state-of-art fabrication (see later figures), is not immediately applicable for competitive commercial OEM use.

Feature size reduction is illustrated in Figure 6.6, from which it will be seen that we are rapidly reaching the limits of optical photolighography.

Figure 6.7a and b show two further aspects of increasing technological ability, which may be read in conjunction with earlier figures. Figure 6.7a indicates that we are beginning to enter the region where system design and testing considerations rather than fabrication limitations are limiting the complexity of systems per chip; increasing CAD expertise will gradually allow the increased fabrication capability to be absorbed where required, after which physical limitations will restrict and control further growth in maximum available chip complexity.

The shrinking of device size which has been the hallmark of the microelectronic industry over the past 20 years therefore may be reaching its end. Device speeds, transistor gains, defect densities, noise immunity, device isolation, and other parameters in silicon technology begin to deteriorate rapidly below the one micron feature size. Dimension sizes approaching 0.1 μm imply that manufacturing tolerances must be down to the level of tens of atoms and tens of electrons, and parasitic capacitances will become a dominant, aggravated by the increasing resistance of thinner interconnect lines.

Alternative technologies such as gallium-arsenide and amorphous silicon [12–19] may not be able to provide very much further impetus in the very-large-scale integrated (VLSI) area: individual gate speeds of 0.1 ns (100 ps) are reported as being commercially available in gallium arsenide (GaAs) medium-scale integration (MSI) by 1985, but to maintain and drive such an operating speed across the chip appears difficult.

6.2.2 Detailed Possible Developments

Within the bounds of foreseeable development, we may note the following features.

Bipolar technology will continue to be developed towards faster switching speeds, but with increasing emphasis upon lower power consumption so as to reduce chip dissipation limitations. Integrated-injection logic (I^2L), integrated-Schottky logic (ISL), and Schottky-transistor logic (STL) (see Chap. 2) may become most acceptable to the custom and semicustom area, but is unlikely to have the "user-friendly" atmosphere of MOS to the OEM designer.

nMOS and CMOS will continue to be scaled down further for some time, possibly bottoming out at around 2 μm feature size for reliable but fast general-purpose semicustom use. Limiting parameters [20] should not be approached for such applications. Bulk-silicon substrate CMOS, which has already experienced the evolution of:

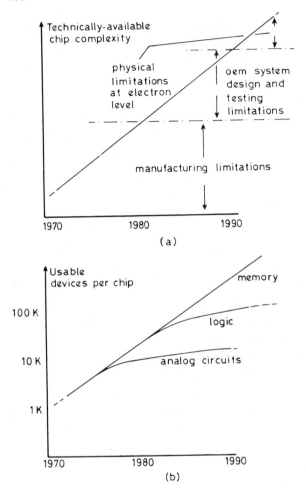

FIGURE 6.7 Future limitations: (a) limiting constraints on chip complexity, (b) usable chip complexity.

Metal-gate, dual guardband isolation, large silicon area per gate,
↓
Metal-gate, single guardband isolation, lower silicon area per gate,
↓
Silicon-gate, oxide isolation, self-aligned gates,

will find chip interconnect topology tackled by:

Two-layer buried polysilicon interconnect, giving denser possible on-chip packing

Two-layer metal interconnect

Two-layer buried low-resistivity polysilicon interconnect, plus one or two-layer metal

Fabrication problems to avoid short-circuits between two-layer metal interconnect have still to be completely solved.

However, it is in the insulated-substrate CMOS area that most improvements should occur. Figure 6.8 indicates the considerable advantages of silicon-on-sapphire-(SOS) CMOS over bulk-silicon substrate CMOS when track capacitance is considered: not apparent in this

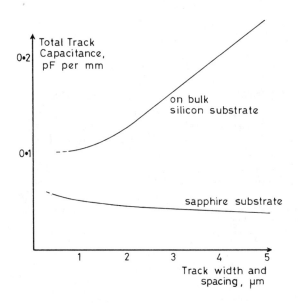

FIGURE 6.8 The interconnect capacitance per track of CMOS, metal running over field oxide, width of metal and spacing between tracks made equal.

illustration is the feature that increasing the number of metallic tracks in parallel increases the total capacitance per track in bulk silicon, but not measurably so with sapphire substrate. Therefore, as has previously been emphasized, since it is not the isolated gate-delay performance which is necessarily the key factor in on-chip operating speed, but rather how fast the gate output can charge and discharge the interconnect capacitance, the reduced track capacitance of insulated-substrate CMOS has powerful advantages.

Other CMOS developments may be in twin-tub V-groove isolation structures, and possibly in newer developments whereby the *p*- and *n*-channel devices are stacked vertically above each other rather than side-by-side on the substrate [21]. If proved to be feasible and reliable, this latter technique will dramatically increase the possible on-chip circuit density.

The precise effect of CAD on custom-specific microelectronics is more difficult to forecast than any other parameter, but its overall effects may be entirely out of proportion to any other development within the next decade. The following may be the decisive effects in the coming years:

1. The sharp divisions between the economics of masterslice gate arrays, standard cell-library, and fully handcrafted design for custom-specific applications will blur.
2. Creation of new cells and macros for a standard cell library will be automatically generated and fully quantified.
3. Generation of test information will be concurrent with the system design, with means to modify the system design as it proceeds in order to improve final system testability.
4. Automatic placement and 100% routing will be available, with built-in design-rule checking.
5. A new area of "soft-cell" library techniques will emerge, wherein the CAD system will tailor the exact silicon layout of the basic building blocks (cells or macros) in order to optimize silicon area, speed, or any other desired parameter for the particular application.
6. The final design data for a given custom-specific application will be held in a standardized fabrication-line-independent data base, the particular fabrication line requirements for mask making or direct-on-wafer writing finally being combined with this fabrication-independent data.

An established (1982) pattern of custom and semicustom design is shown in Figure 6.9 [22]. The standard cell library would be of polycell design with standardized cell height, while the soft-cells would be handcrafted variants of the standard polycells. Fully handcrafted design would be at individual transistor level. What the above CAD

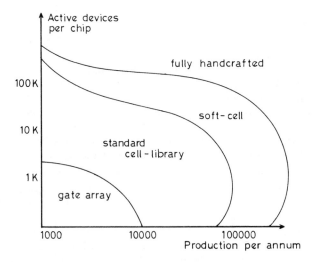

FIGURE 6.9 An existing subdivision of custom design techniques for optimum unit cost of product. (Data from Silicon Microsystems, Malmesbury, England.)

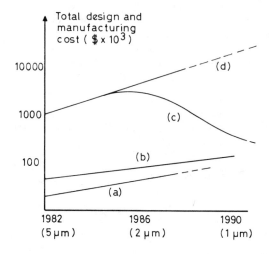

FIGURE 6.10 Estimated trends in design and manufacturing costs for the supply of 10,000 state-of-the-art circuits: (a) gate array, 2000-gate chip in 1982, 10,000-gate chip in 1990; (b) standard cell, 3000-gate chip in 1982, 50,000-gate chip in 1990; (c) full custom using cell-library macros or soft-cells, up to 100,000 gates in 1990; (d) fully handcrafted versions of (c), possibly never done after mid 1980s.

possibilities imply is that the boundaries between fully handcrafted, soft-cell, and standard cell-library would merge, leaving a continuous design capability which offers both optimum circuit density and performance without excessive design costs. Masterslice gate arrays, however, would probably still exist, particularly for very rapid prototype and research and development purposes, where the cost of a complete mask set for a few circuits would not be justified [23].

Figure 6.10 illustrates another current forecast of the future which may require modification should more powerful CAD tools become available quicker than presently anticipated. The tendency in this area would be to reduce costs and to make the breakeven production quantity for a custom circuit become considerably lower than now forecast.

6.3 SUMMARY

The underlying theme of this text has been the use by OEM designers of the extensive and still growing expertise in microelectronic fabrication. Custom-specific microelectronics is still in its formative years, but we will witness its application worldwide within the coming decade, with the slow decline in prominence of the standard off-the-shelf products.

The majority of applications for microelectronics are not large products and do not have a very great glamour factor. However, the present objective of large system designers and research and development teams is the completion of the following evolutionary trail, started less than three decades ago, and hopefully to be achieved within the coming decade:

individual discrete semiconductor components; pc board assemblies (1960s)

individual SSI/MSI packages; pc board assemblies (1970s)

individual LSI/VLSI packages, pc board assemblies (1980s)

WAFER-SCALE INTEGRATION

LSI/VLSI complexity "parts" on the same wafer, with interconnections on the wafer to form a complete large working system

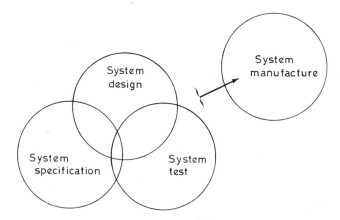

FIGURE 6.11   Wafer-scale integration, full system on one wafer, with system specification, design, and test as an overlapping part of the design-on-silicon.

Hence, complete systems-on-silicon as indicated in Figure 6.11 is the objective, thus maximizing the number of on-chip system connections, and minimizing the total number of input/output (I/O) connections to be made to silicon from other parts.

The continuing spinoff from these research activities will undoubtedly be to the benefit of the more humble products, both in terms of availability and cost as well as reliability and commercial acceptance.

REFERENCES

1. Morrow, J., and Victor, J. Looking ahead: trends and forecasts. *EDN* 28(10):304, 1983.
2. Penisten, G. E. A survey of the present and future status of the semi-custom arena. *J. Semicustom ICs*, 1(2):5—11, 1983.
3. Summers, T. Logic arrays or standard cells? An objective view of the semicustom business. *Proc. 3rd Int. Conf. Semi-Custom ICs*, paper 2/10, London 1983.
4. Rappaport, A. The semicustom IC future: user's perspectives. *Proc. 3rd Int. Conf. Semi-Custom ICs*, paper 1/11, 1983.
5. Special Report. Semicustom circuits. *Electronic Times*, 23—38, August 18, 1983.
6. Special issue. Microelectronics in GEC. *GEC J. Science Technology*, 48(2):1982.

7. Source III, Inc. *Gate Arrays: Implementing LSI Technology.* Source III, Inc., San Jose, CA, 1982.
8. Arnold, J. A. Technology and design aids. *Proc. IEE 2nd Int. Conf. on the Impact of VLSI Technology on Communication Systems,* 47—49, 1983.
9. Szirom, S. Z. Custom-semicustom IC business report. *VLSI Design,* 4(1):32—38, 1983.
10. Mackintosh Profile Report. *Profile of the Worldwide Semiconductor Industry.* Benn Electronic Publications, Luton, U.K., 1982.
11. Krejcik, M., and Mash, S. (Eds.). *Semi-custom IC Year Book 1983.* Benn Electronic Publications, Luton, U.K., 1983.
12. Muroga, S. *VLSI System Design.* Wiley-Interscience, New York, 1982.
13. Holmes, L. Amorphous-silicon devices begin to shape up. *IEE Electronics Power,* 29:222—225, 1983.
14. DiLorenzo, J., and Khandelwal, K. (Eds.). *GaAs FET Principles and Technology,* Artech House, New York, 1981.
15. Solomon, P. M. A comparison of semiconductor devices for high-speed logic. *Proc. IEEE,* 70:489—509, 1982.
16. Nishi, Y. Comparison of new technologies for VLSI: possibilities and limitations. *J. Microelectronics,* 12(6):5—10, 1981.
17. Capace, R. P. Special Report: new lsi processes. *Electronics,* 52(19):109—115, 1979.
18. Eden, R. C., Livingston, A. R., and Welch, B. M. Integrated circuits: the case for gallium arsenide. *IEEE Spectrum,* 20(12): 30—37, 1983.
19. Mellor, P. J. T. Gallium arsenide digital integrated circuits. *British Telecom Technology,* 2(1):60—68, 1984.
20. Richer, J. Electromigration design limits in CMOS. *Proc. 3rd Int. Conf. Semi-Custom ICs,* paper 3/7, London 1983.
21. Davis, R. K. The case for CMOS. *IEEE Spectrum,* 20(10):26—32, 1983.
22. Cambell, J. Rapid design times with CAD of standard cell libraries and gate arrays. *Proc. 2nd Int. Conf Semi-Custom ICs,* paper 2/7, London 1982.
23. Balch, J. W., Wayne-Current, K., Magnuson, W. G., and Pocha, M. D. Introducing gate arrays to engineers. *J. Semi-custom ICs,* 1(2):13—18, 1983.

Appendix/A
A Review of Semiconductor
Fundamentals and Charge Carriers

To supplement the device details given in Chapter 2, we cover here in a very brief form, the basic concepts of semiconductor materials and charge carriers. This review will be largely descriptive and nonmathematical, but hopefully instructive for those who are not intimately concerned with device physics. For more in-depth treatments, reference may be made to many detailed texts [1-10].

The atomic model of the atom is well known. In its simplest form it is considered to consist of a central nucleus containing the major mass of the atom, which is positively charged, surrounding which are the orbits of electrons which with the nucleus comprise the complete atom. The circulating electrons, one in the case of the hydrogen atom, two in the case of helium, and so on, circulate in well-defined orbits, being grouped into distinct bands or shells which can contain up to a given maximum number of electrons per shell. In the case of germanium and silicon we have the atomic structures shown in Figure A.1.

The outer shell of all atoms contains the valency electrons, and gives the atom its normal chemical and electrical properties. When all the electrons in this outer shell are firmly retained in orbit around the nucleus, the atom is an electrical isolator; however, when valency electrons are more loosely bound, the material becomes a conductor due to the movement and interchange of the negatively charged electrons. The chemical properties are also controlled by the number of these outer valency electrons, as will be recalled from elementary chemistry. In all normal circumstances we may discount the presence of all electrons in the shells below the outer valency electrons, and consider them and the nucleus as the unchanging central core structure of the atom.

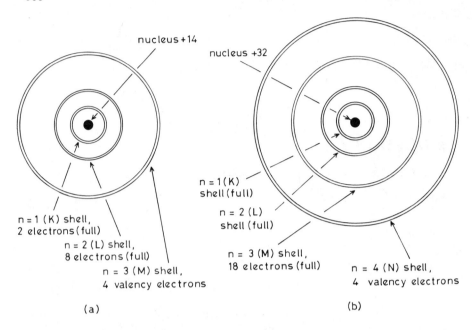

nucleus +14

nucleus +32

n = 1(K)
shell (full)

n = 2 (L)
shell (full)

n = 1 (K) shell,
2 electrons (full)

n = 2 (L) shell,
8 electrons (full)

n = 3 (M) shell,
4 valency electrons

n = 3 (M) shell,
18 electrons (full)

n = 4 (N) shell,
4 valency electrons

(a)   (b)

FIGURE A.1 The simplified atomic model of an isolated silicon or germanium atom (radii not to scale): (a) silicon, atomic weight 14 (2,8,4), (b) germanium, atomic weight 32 (2,8,18,4). Note that the allowed atomic shells may be subdivided into further allowable subshells.

Quantum mechanics defines that the orbiting electrons of atoms can only have certain values of total momentum and that the transfer of an electron from one orbit to another orbit of larger radius can only occur when the electron receives appropriate energy to allow it to make such a quantum jump. Similarly, the movement of an electron back from a higher orbit to a lower radius one must be accompanied by quantized release of energy, in the form of light or some other frequency of radiation. Additionally, by the Pauli exclusion principle, only one electron is allowed to occupy a specific energy level in an atom at any one time. From these laws we may attempt to describe the atomic and molecular characteristics of conductors, nonconductors (insulators), and semiconductors, with our specific interest being in the latter.

Let us first concentrate on the atomic structure of silicon, from which we may generalize to germanium and other semiconductor compounds as well as to conductors, and nonconductors.

As previously noted, the outer electron shell of an isolated silicon atom contains four electrons. This M shell (see Fig. A.1) is divided into three possible subshells which may contain two, six, and ten electrons, respectively, and hence at its lowest energy state (at 0°K) the silicon atom has two electrons in each of the two inner M subshells. However, when atoms are in close proximity to each other a modification in the energy states of individual atoms arises, since interaction between the outer electron orbits can occur. This is particularly true when the atoms are arranged in a perfect molecular crystalline structure, as is possible in the case of silicon and other quadravalent materials.

The modification in the energy states of the outer electrons when going from the picture of individual isolated silicon atoms to a crystalline structure is that exactly one half of the total atomic energy states of the valency electrons are depressed and one half of the total are increased, the overall kinetic energy of the electrons in the molecular assembly remaining unchanged. In total there arises a spread in allowed energies in contrast to the specific electron energies allowable in the isolated atom model, forming two allowable bands of energy as illustrated in Figure A.2.

The lower band of allowable energy states is termed the "valency band," above which is the "forbidden band" or "band gap" of energies in which no electrons can exist. Above this is the outer "conduction band." The allowable states in the valency band are such that all four valency electrons per atom can be accommodated, and hence, at 0°K the valency band is completely full with no electrons being present in the outer conduction band. Under these perfect minimum-energy conditions all electrons would be held in the crystal lattice structure, all being retained in the inner valency band, and the material would be a perfect

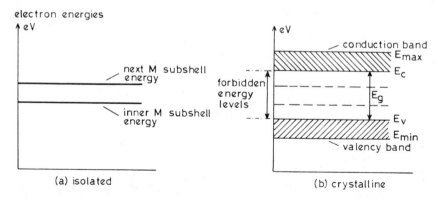

(a) isolated   (b) crystalline

FIGURE A.2   The electron energy model for silicon: (a) electron energies of the valency electrons present in isolated individual Si atoms, all atomic energies identical with precisely defined energy levels, (b) valency electron energies present in perfect crystalline Si, with a spread of energies throughout the crystal into two permitted energy bands.

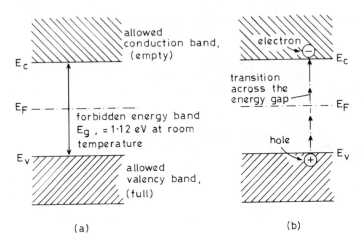

(a)   (b)

FIGURE A.3   The energy band model for crystalline silicon: (a) perfect crystalline Si at 0°K, no valency electrons in the conduction band,
(b) the transition of an electron across to the conduction band on receipt of energy E_g, resulting in the generation of an electron-hole pair
for conduction pruposes. Note that the energy level E_F is the Fermi
energy level for intrinsic crystalline silicon (see Ref. [3] or other
references).

insulator since no electrons would be available for electrical transport
through the material in response to an applied electric field.

In this energy band model the highest valency band energy is E_V,
while the lowest conduction band energy is E_c. The band energy gap
E_g is therefore simply

$$E_g = (E_c - E_v)$$

all energies normally being expressed in electron volts.* Thus, in order for a material to become an electrical conductor, electrons have to
be made available in the conduction band, which in the case of crystalline silicon entails some electrons being given the additional energy E_g
to jump across into the conduction band and hence become available as
conduction carriers. Heat and incident radiation is sufficient to cause
a small number to acquire this additional energy, and therefore we have

----

*One electron volt is the energy acquired by an electron when accel-
erated through a voltage difference of 1 volt, and is equal to 1 eV =
(1 volt × charge on an electron in coulombs), = 1.6×10^{-19} joules.

TABLE A.1   The Forbidden Energy Gaps of Various Semiconductor Elements and Compounds

Material	Symbol	Energy gap eV
Germanium	Ge	0.72
Silicon	Si	1.12
Diamond	C	6.70
Indium phosphide	InP	1.25
Gallium-arsenide	GaAs	1.40
Gallium-phosphide	GaP	2.30
Cadmium telluride	CdTe	1.50
Cadmium selenide	CdSe	1.70
Lead telluride	PbTe	0.3
Lead selenide	PbSe	0.3

the intrinsic semiconducting properties of the material, being extremely
poor in comparison with the conductivity of metals. This model is illustrated in Figure A.3 which also shows that when an electron is excited
into the conduction band it leaves behind a vacancy or "hole" in the
valency band, and the presence of electron-hole pairs is inherent in the
intrinsic conductivity of the material.

The values of E_g for different semiconductor materials are given in
Table A.1. Note that the very high value of 6.7 eV for diamond means
that for all practical purposes this is an almost perfect insulator. Indeed
as may now be appreciated, it is the value of the energy band gap which
is the distinction between conducting, semiconducting, and insulating
materials: the wider the gap (the higher the value for E_g) then the
poorer is the material as an electrical conductor and vice versa. The
band gap in metals is extremely low, or indeed, may not even exist due
to an overlap between E_C and E_V: in such cases there are always plenty
of electrons available for conduction purposes and the electrical resistivity ρ is consequently low (conductivity $\sigma = 1/\rho$ = high). On the other
hand, silicon dioxide, which is the principal insulating material in semiconductor fabrication technology, has a value E_g of about 8 eV, and
consequently is an extremely efficient insulator. It will also be noted
that with semiconductor materials where there exists a given band gap
E_g, an increase in temperature will excite more electrons into the conduction band, and hence reduce the intrinsic resistivity of the material.
This negative resistance/temperature characteristic is the cause of

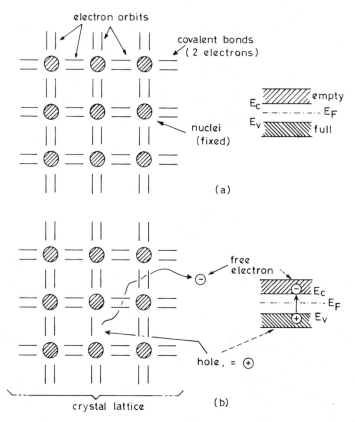

FIGURE A.4 The covalent lattice bonding model of crystalline silicon: (a) the perfect 0°K mode, all nuclei bound together by the four valency electrons per atom. (b) The result of breaking one covalent bond, releasing one electron from the crystal lattice and creating one hole.

semiconductor problems at increasing ambient temperature conditions, and must be accommodated by appropriate circuit design.

The energy band model of semiconductors may be supplemented with the familiar molecular bonding picture of quadravalent atoms. Silicon and other similar atoms may be assembled in a diamond lattice molecular structure wherein each atom exhibits a bonding to its four nearest neighbors in three-dimensional space. This bonding involves the mutual sharing of the four valency electrons, giving covalent bonds involving two valency electrons between every pair of atoms. When the crystal structure is perfect and all covalent bonds are intact, then the four valency electrons of each atom freely orbit around any adjacent nuclei, as indicated in the two-dimensional representation shown in Figure A.4a. This "sharing" of electrons between nuclei spreads the allowable energy levels for the electrons from the precisely defined orbits of an isolated atom to form the allowable valency and conduction energy bands which we have previously considered.

If we assume that the crystal lattice structure is perfect, and that all atoms are correctly positioned with respect to each other, then at 0°K all covalent bonds will be intact and no valency electrons will be available for conduction purposes. At higher temperatures, however, certain bonds will be broken by incident energy, and electrons will be free for conduction purposes. This excitation state is shown in Figure A.4b, and corresponds to our energy band model of Figure A.3. Note that the energy necessary to break a bond in this molecular model is just another way of expressing the energy necessary to jump the forbidden energy gap of the energy band model.

Both the electrons and the resulting holes are free to take part in the conduction process. Recall that the hole may be regarded as a charge carrier in its own right, although in reality it is a drift of negatively charged electrons from intact covalent bonds into vacant bond positions, leaving behind holes now in a different place in the crystal lattice.* However, the mobility, that is, the ease with which carriers may move under the influence of an electric field, is different for holes and unbound (free) electrons. The parameter of charge carrier mobility, μ, is defined as

$$\mu = \frac{\text{Average value of carrier drift velocity in m/s}}{\text{electric field strength in V/m}}$$

and will be lower for holes than for electrons. Typical semiconductor values are

Semiconductor	Drift current mobility μ (m^2 V^{-1} s^{-1})	
	μ_n(Electrons)	μ_p(Holes)
Germanium	0.39	0.19
Silicon	0.15	0.05
Gallium-arsenide	0.5	0.03
Gallium-phosphide	0.01	0.002

*Recall that in the absence of any applied electric field there is normally a random movement of electrons through the crystal structures as holes and electrons change position. This random movement is known as the diffusion current. However when an electric field is present, then the electrons tend to migrate to the positive pole, and the holes toward the negative pole, this movement of charge carriers being termed draft current.

Unfortunately, many semiconductor materials which possess the desirable characteristic of high charge carrier mobility have very low bandgap E_g values, which make them unsuitable for practical devices. Hence, there is usually conflict between desirable E_g and μ values.

So far we have merely considered the intrinsic conduction in silicon, assuming that the silicon contains an insignificant number of impurity atoms, and that the molecular crystal lattice structure is perfect. The situation for germanium and other semiconductor materials is fundamentally similar, and needs no further exposition in this brief review. However, intrinsic semiconductor properties are of little practical value except in certain positive-temperature-coefficient thermistor devices, but instead it is the ability to add controlled amounts of impurity levels to semiconductor materials in order to modify their energy band levels and carrier availabilities which gives us the properties required for microelectronic use. As a measure of the relatively few intrinsic conduction carriers present in the conduction band in germanium and silicon, the values are approximately:

Germanium: 10^{13} electrons/cm^3 at room temperature (and the same number of holes)

Silicon: 10^{10} electrons/cm^3 at room temperature (and the same number of holes)

which for silicon means that only about 1 covalent bond in 10^{13} is broken at room temperature, since there are about 5×10^{22} atoms/cm^3 each with four valency electrons in crystalline silicon. For germanium, with its lower value of E_g and larger atomic size, there is a somewhat higher percentage of broken bonds, but the percentage is still minutely small.

The addition of a controlled amount of specific impurity atoms into the crystalline lattice structure (doping) increases either the electron or the hole-carrier concentration. To increase the electron carrier concentration, the addition of pentavalent atoms, that is group V atoms with five valency electrons, is made: to increase the hole-carrier concentration, the addition of trivalent atoms, that is group III atoms with three valency electrons, is made. Pentavalent dopants for silicon ("donor" impurities) include phosphorus, antimony, or arsenic, phosphorus being most common: trivalent dopants ("acceptor" impurities) include boron, gallium, or indium, boron being most common. Doping concentrations of about 1 in 10^8 will clearly modify the electrical characteristics of the material considerably from its intrinsic performance.

The molecular bonding model previously introduced in Figure A.4 now becomes modified as shown in Figure A.5. Each donor impurity atom now provides one electron which is free from the covalent bonding requirements, while each acceptor impurity atom provides one hole, additional to the intrinsic molecular model. Thus, additional negative

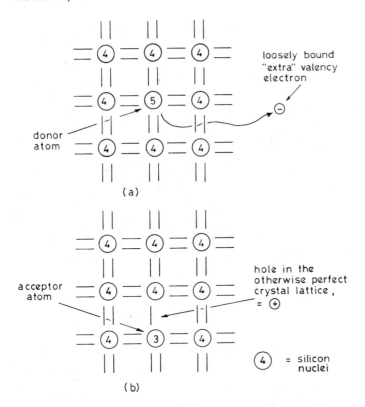

FIGURE A.5   The effect of a donor and an acceptor impurity atom in
the crystalline silicon bonding model. (a) Donor atom, say phosphorus,
with its fifth valency electron which is superfluous to the quadravalent
crystal lattice structure. (b) Acceptor atom, say boron, with its three
electrons which gives an electron deficit (positive hole) in the lattice
structure.

(electron) and positive (hole) carriers in controlled amounts can be
provided.   The donor impurities provide n-type semiconductor material,
while acceptor impurities provide p-type semiconductor material.   However note that

1.   The increase in negative electron carriers in n-type material does
     not increase the hole-carrier density; the majority carriers are
     therefore electrons, holes being minority carriers.
2.   The increase in positive hole carriers in the p-type material does
     not increase the electron carrier density; the majority carriers
     are therefore holes, electrons being minority carriers.

3. The higher carrier mobility μ of electrons compared with holes
    means that n-type materials should have a faster performance
    than corresponding p-type materials.

   The effect of introducing donor and acceptor impurities into the
silicon crystalline structure can also be considered in the energy band
model. The effect of donor impurity atoms is to create an energy level
very close to the conduction band, in which the "extra" electron per
impurity atom would exist at 0°K. Electrons can therefore easily leave
this level and appear as mobile carriers in the conduction band. Similarly, the effect of acceptor impurity atoms is to create an empty energy level very close to the valency band, into which electrons from the
valency band can easily jump, thus creating additional mobile holes in
the valency band. This is shown in Figure A.6. The binding energies
of various donors and acceptors are given below, which means that with
such small energy gaps virtually all the electrons are in the conduction
band at room temperature, and similarly all holes are present in the
valency band:

Group V donors	$\|E_c - E_d\|$	Group III acceptors	$\|E_a - E_v\|$
Phosphorus (P)	0.044 eV	Boron (B)	0.045 eV
Antimony (Sb)	0.039 eV	Gallium (Ga)	0.065 eV
Arsenic (As)	0.049 eV	Indium (In)	0.16 eV

   The above nonmathematical discourse should give sufficient introductory insight into the general conductivity characteristics of n- and
p-type material for most engineering purposes. Note that any discon-
tinuities in the crystal lattice structure of the material will have pro-
found effects, as will impurities in the material before selective doping
and therefore, scrupulous attention to manufacturing purity is essential.
However, we should continue one stage further in this general discussion
in order to consider the effects of adjacent p- and n-type doping within
the same crystal lattice, that is the pn junction.
   The operation of a pn junction can best be described by a continua-
tion of our previous energy band model. Figure A.7a and b shows the
energy band model for separate p- and n-type materials as before.
However, when these two doping concentrations occur within *the same*
crystal lattice there is a diffusion of the free electrons from the conduction band of the n-type material into the p-type region, and similary
a diffusion of free holes from the p-type material into the n region.
This diffusion of electrons from the n-type region and holes from the

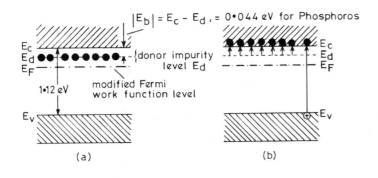

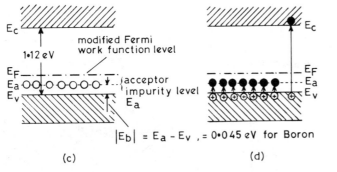

FIGURE A.6 The electron energy band model for crystalline silicon with donor and acceptor impurities giving n- and p-type material, respectively. (a) Donor impurities, donors' pentavalency results in occupied electron level E_d at 0°K. (b) At room temperature, virtually all donor impurity electrons now in conduction band, plus a small minority of intrinsic conduction carriers (see Figure A.3b); majority carriers = electrons. (c) Acceptor impurities, acceptors' trivalency gives unfilled electron energy level E_a at 0°K. (d) At room temperature energy level E_a virtually filled, leaving holes as carriers in the valency band; majority carriers = holes.

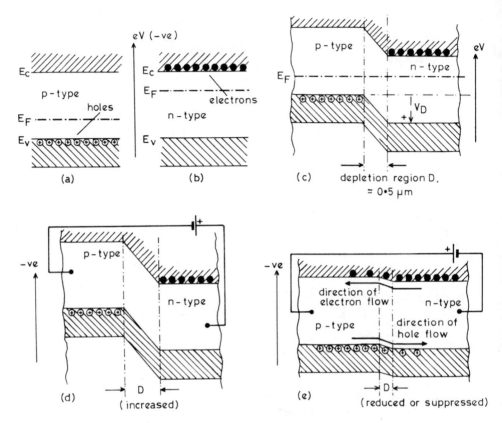

FIGURE A.7 The electron energy band model for a simple *pn* junction:
(a) the previous *p*-type model (Fig. A.6d) ignoring minority carriers,
(b) the previous *n*-type model (Fig. A.6b) ignoring minority carriers,
(c) the *pn* junction model in equilibrium with no applied voltage, giving
depletion region D and junction potential difference V_D, (d) the effect
of an applied voltage, positive to *n*-type region (reversed-biased junction), (e) the effect of an applied voltage positive to the *p*-type region
(forward-biased junction).

p-type region however causes the n-type region to become positively charged with respect to the p-type region, and equilibrium conditions will be reached when the potential gradient between the two sides is sufficient to stop any further carrier diffusion across the boundary.[*] This is illustrated in Figure A.7c.

Recombination and annihilation of positive and negative carriers occur either side of the boundary, leaving a space-charge or "depletion layer" region at the boundary, in which no free holes or electrons can permanently exist. Exact equilibrium conditions are when the Fermi energy level of the two materials reaches a common value, which will be when the potential gradient or "barrier potential" is about 0.7 V for silicon or 0.3 V for germanium.

Figures A.7d and e now illustrate the effect of applying a voltage across the pn junction. With positive voltage to the n-type region, the effect is to increase the barrier potential across the junction: after an initial transitory diffusion current flow which increases the width of the depletion layer, equilibrium conditions again are established, with ideally zero current flow. This is the pn junction in reverse bias. On the other hand, positive voltage to the p-type region will reduce the barrier potential, and appreciable current will flow when the applied potential is sufficient to eliminate the potential gradient. This is the pn junction in forward bias.

This is a major simplification of the full details of the pn junction characteristics, but with the previous data it is hopefully sufficient for the purposes of this text. For more specific and detailed information, including the necessary distinctions between drift current, diffusion current, and generation/recombination current and more precise considerations of electron energies see the following references or many other professional publications.

REFERENCES

1. Adler, R. B., Smith, A. C., and Longini, R. L. *Introduction to Semiconductor Physics*. John Wiley, New York, 1964.
2. Streetman, B. G. *Solid-state Electronic Devices*. Prentice-Hall, Englewood Cliffs, NJ, 1980.

---

[*]The diffusion of each electron out of the n-type region will leave behind a nucleus whose positive charge is no longer balanced by the electron. Hence, the n-type material will have a unbalance of positive and negative charges, with excess positive charge. Conversely for the holes, each of which will leave behind a net negative charge.

350 *Appendix A*

3. Pierret, R. F. *Semiconductor Fundamentals*. Modular Series on Solid-state Devices, Vol. 1. Addison-Wesley, Reading MA, 1983.

4. Grove, A. S. *Physics and Technology of Semiconductor Devices*. John Wiley, New York, 1967.

5. Sze, S. M. *Physics of Semiconductor Devices*. Wiley-Interscience, New York, 1969.

6. Moll, J. L. *Physics of Semiconductors*. McGraw-Hill, New York, 1964.

7. Glaser, A. B., and Subak-Sharpe, G. E. *Integrated Circuit Engineering*. Addison-Wesley, Reading MA, 1977.

8. Talley, H. E., and Daugherty, D. G. *Physical Principles of Semiconductor Devices*. Iowa State University Press, Ames, Iowa, 1976.

9. Pohl, H. A. *Quantum Mechanics for Science and Engineering*. Prentice-Hall, Englewood Cliffs, NJ, 1967.

10. Seymour, J. *Physical Electronics*. Pitman, London, 1972.

Appendix/B
Further Supporting Data

To supplement the more specific information covered in the main chapters of this text, we give below some further information which may
be of supporting interest.

Tables B.1 and B.2 give the atomic structure of the chemical elements in the form of the familiar periodic table, with particular emphasis upon (1) those elements which have four electrons in their valency
band, and (2) those elements in the immediately adjoining valency bands
with fewer or more valency electrons. From Table B.2 it may be noted
that silicon and germanium are the second and third listed elements in
group IV. The first element in this group, namely carbon (diamond),
has too high an energy gap E_g to be suitable as a semiconductor material: it is essentially an insulator. The higher-placed group IV elements,
however, namely tin and lead, have too low an energy gap, and are
effectively conductors.

Thus, of all the basic chemical elements only silicon and germanium
are appropriate for useful semiconductor purposes. However, further
semiconductor *compounds* may be possible, particularly compounds made
from group III and group V such as gallium-arsenide, indium phosphide,
and others. (See also Table A.1 in the Appendix A). This work on
III/V and other possible compounds is the subject of much ongoing research activity at this fundamental device level.

Tables B.3 and B.4 give details of the electrical conductivity and
resistivity of materials, including the effect of donor and acceptor impurities in crystalline silicon. The exact range of resistivity values
which one would normally refer to as the semiconductor region is

TABLE B.1 A Selection from the Periodic Table of
the Elements, in Ascending Atomic Number (Total Electrons per Atom)

Element, Symbol and Atomic Number	Electron Shells and Occupancy						
	K	L	M	N	O	P	Q
Hydrogen, H, 1	1						
Helium, He, 2	2						
Lithium, Li, 3	2	1					
Beryllium, Be, 4	2	2					
Boron, B, 5	2	3					
Carbon, C, 6	2	4					
Nitrogen, N, 7	2	5					
Oxygen, O, 8	2	6					
Fluorine, F, 9	2	7					
Neon, Ne, 10	2	8					
Sodium, Na, 11	2	8	1				
Magnesium, Mg, 12	2	8	2				
Aluminium, Al, 13	2	8	3				
Silicon, Si, 14	2	8	4				
Phosphorus, P, 15	2	8	5				
Sulphur, S, 16	2	8	6				
Chlorine, Cl, 17	2	8	7				
Argon, A, 18	2	8	8				
Potassium, K, 19	2	8	8	1			
Calcium, Ca, 20	2	8	8	2			
Scandium, Sc, 21	2	8	9	2			
Titanium, Ti, 22	2	8	10	2			
Vanadium, V, 23	2	8	11	2			
Chromium, Cr, 24	2	8	13	1			
Manganese, Mn, 25	2	8	13	2			
Iron, Fe, 26	2	8	14	2			
Cobalt, Co, 27	2	8	15	2			
Nickel, Ni, 28	2	8	16	2			
Copper, Cu, 29	2	8	18	1			
Zinc, Zn, 30	2	8	18	2			
Gallium, Ga, 31	2	8	18	3			
Germanium, Ge, 32	2	8	18	4			
Arsenic, As, 33	2	8	18	5			
Selinium, Se, 34	2	8	18	6			
Bromine, Br, 35	2	8	18	7			
Krypton, Kr, 36	2	8	18	8			
⋮							
Silver, Ag, 47	2	8	18	18	1		
Cadmium, Cd, 48	2	8	18	18	2		
Indium, In, 49	2	8	18	18	3		
Tin, Sn, 50	2	8	18	18	4		
Antimony, Al, 51	2	8	18	18	5		
⋮							
Gold, Au, 79	2	8	18	18	32	1	
Mercury, Hg, 80	2	8	18	18	32	2	
Thallium, Tl, 81	2	8	18	18	32	3	
Lead, Pb, 82	2	8	18	18	32	4	
Bismuth, Bi, 83	2	8	18	18	32	5	
⋮							
Uranium, U, 92	2	8	18	32	18	12	2

TABLE B.2 A Selection from Table B.1, Giving the Atomic Elements with Complete Shells Below the Final Outer (Valency) Shell, Outer Shell from One (Minimum) to Eight (Maximum) Valency Electrons. The Atomic Number Is Given in Parentheses

GROUP	I	II	III	IV	V	VI	VII	VIII
1 shell	H(1)	–	–	–	–	–	–	He(2)
2 shells	Li(3)	Be(4)	B(5)	C(6)	N(7)	O(8)	F(9)	Ne(10)
3 shells	Na(11)	Mg(12)	Al(13)	Si(14)	P(15)	S(16)	Cl(17)	A(18)
4 shells	Cu(29)	Zn(30)	Ga(31)	Ge(32)	As(33)	Se(34)	Br(35)	Kr(36)
5 shells			In(49)	Sn(50)	Sb(51)			
6 shells			Tl(81)	Pb(82)	Bi(83)			
7 shells*								

```
potential              potential              potential
p-type dopants         semiconductor          n-type dopants
(acceptor impurities)  elements in            (donor impurities)
in this category,      this category,         in this category,
3 valency electrons    4 valency electrons    5 valency electrons
```

* 7-shell atomic structures contain the unstable radioactive elements

difficult to define precisely: possibly from $\rho = 0$ to $\rho = 10^{-8}$ ohm/cm may be regarded as the "semiconductor" region.

In Tables B.5 and B.6 we revert back again specifically to micro-electronics fabrication. Table B.5 gives the general stages involved in normal photolithographic processing of wafers: exact steps may vary from manufacturer to manufacturer and technology to technology, but the basic flow will be generally as indicated. The final table shows the possible means of going from the pattern-generation design data for the masking, invariably held on magnetic tape in some prescribed data format, to the photographic plates or other means of imprinting the masking patterns on the wafer surface. The most usual method at the present time is a ×1 working plate with direct (or near direct) contact of the working plate on the wafer. This procedure is well established but resolution is limited, and working plate life is limited due to mechanical handling, scratches, and pick-up problems. Hence, optical or other beam projection methods are finding increased favor, and indeed will become necessary when feature sizes on the wafer are reduced to below the 3 μm level.

However, while the circuit and equipment designer should be aware of the factors involved in the final fabrication of the custom or semi-custom circuits he may be designing, the continuously evolving details of fabrication will remain the province of fabrication line staff as long as can be envisaged.

TABLE B.3 The Electrical Resistivity ρ and Conductivity σ of Various Materials at Room Temperature $\rho = \Omega/\text{cm}$, $\sigma = 1/\rho = \Omega/\text{cm}$

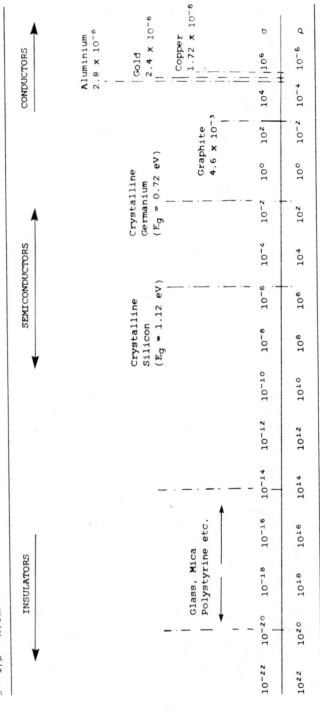

TABLE B.4 The Variation of Electrical Resistivity ρ for Silicon at Room Temperature as a Function of Impurity Concentration Crystalline Silicon $\simeq 5 \times 10^{22}$ atoms/cm^3

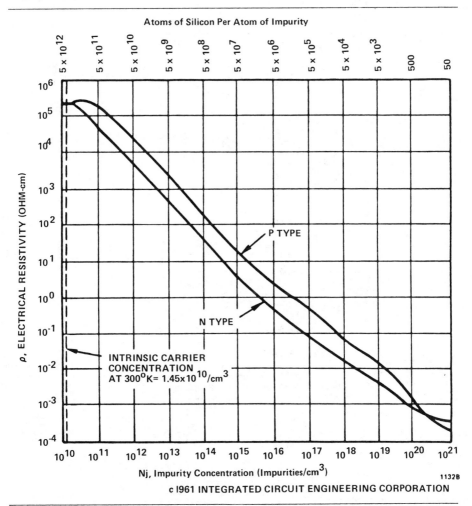

Source: Reproduced by courtesy of Integrated Circuit Engineering Corporation, Scottsdale, Arizona; their publication No. 10665-01C, p. 51, 1980.

TABLE B.5 The Steps in Conventional Photolithography Process
Stages

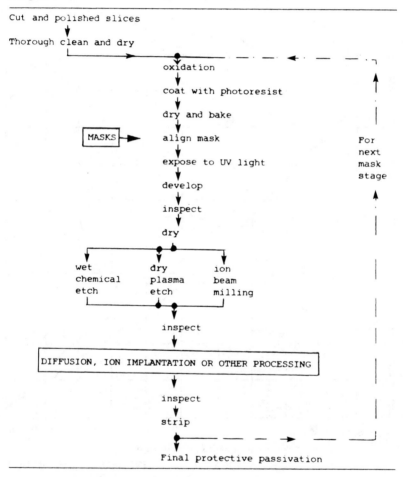

Cut and polished slices

Thorough clean and dry

oxidation

coat with photoresist

dry and bake

MASKS → align mask

expose to UV light

develop

inspect

dry

wet chemical etch   dry plasma etch   ion beam milling

inspect

DIFFUSION, ION IMPLANTATION OR OTHER PROCESSING

inspect

strip

Final protective passivation

For next mask stage

TABLE B.6 Lithography Techniques for Each Mask Stage from the Design Data on the Pattern-Generation Magnetic Tape.[a]

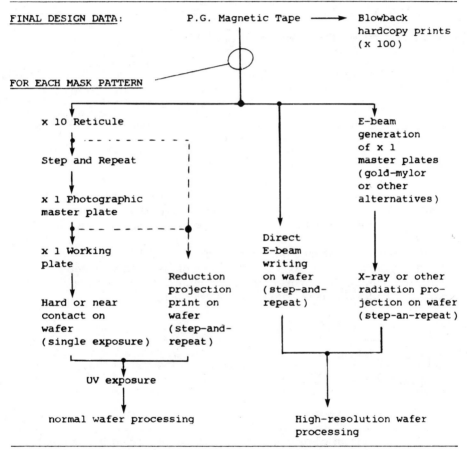

[a]Note: × 10, × 1, etc., refer to the size compared with the final chip dimensions.

Appendix/C
The Number of Dies Per Wafer

We finally list below some dimensional and statistical data giving:
(1) the approximate number of dies (chip layouts) of increasing size
which can be laid down by step-and-repeat on different diameter wafers
and (2) die yield as a function of die size with current fabrication expertise.

The tabulation of dies per wafer is given in Table C.1 being tabu-
lated for die size in microns versus wafer diameter in centimeters, and
assuming square dimensional dies. For chip size in mils (thousands of
an inch) and wafer diameter in inches, recall that

1 micron = 0.04 mils

1 mil = 25.4 microns

1 mil^2 = 645 square microns

1 inch = 2.54 centimeters

The final graph of probe yield versus die size given in Figure C.1
assumes yield is independent of wafer size, which may not be entirely
correct. Generally, increasing expertise in fabrication will tend to
make defect density lower, and hence, yields higher than shown for
wafer sizes for example up to 4 in. (10.1 cm) diameter. The yield
per wafer is given by the statistical equation

TABLE C.1   Number of Whole Dies Possible Versus Wafer Diameter

Square die size (μm x μm)	Die area (μm^2)	Approximate no. of die per wafer			
		wafer size (diameter)			
		5 cms (2")	7.6 cms (3")	10.1 cms (4")	12.7 cms (5")
1000	1×10^6	1880	4320	7730	1900
1250	1.56×10^6	1200	2760	4940	7550
1500	2.25×10^6	805	1880	3490	5200
1750	3.06×10^6	580	1370	1520	3820
2000	4×10^6	435	1035	1840	2920
2250	5.06×10^6	340	835	1480	2290
2500	6.25×10^6	275	655	1180	1800
2750	7.56×10^6	220	545	980	1490
3000	9×10^6	188	450	820	1250
3250	10.56×10^6	156	372	703	1090
3500	12.25×10^6	133	324	601	935
3750	14.06×10^6	112	284	510	810
4000	16×10^6	97	256	448	710
4250	18.06×10^6	85	214	400	625
4500	20.25×10^6	76	190	350	550
4750	22.56×10^6	67	168	313	490
5000	25×10^6	61	148	282	445
5250	27.56×10^6	52	135	264	395
5500	30.25×10^6	49	123	230	360
5750	33.06×10^6	43	110	212	316
6000	36×10^6	39	98	190	300
6250	39.06×10^6	33	90	172	273
6500	42.25×10^6	32	82	158	250
6750	45.56×10^6	30	75	146	232
7000	49×10^6	24	68	134	217
7250	52.56×10^6	22	67	124	197
7500	56.25×10^6	22	60	115	184
7750	60.63×10^6	22	52	105	173
8000	64×10^6	18	52	100	162
8250	68.06×10^6	16	51	90	152
8500	72.25×10^6	16	43	86	142
8750	76.56×10^6	16	42	81	132
9000	81×10^6	15	40	80	125

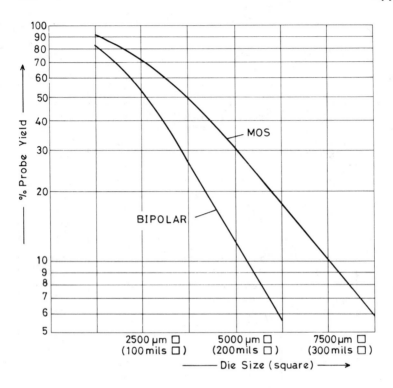

FIGURE C.1   Statistical wafer probe yield as a function of die size
(see text).

$$\gamma = \left\{ \frac{1 - e^{-(A/A_0)}}{A/A_0} \right\} \times 100\%$$

where A is area of die, A_0 is wafer area per defect, and where A_0 for
bipolar technology may be taken as 1×10^7 μm^2/defect (15000 $mils^2$/de-
fect, and for typical metal-oxide semiconductor technology as 2×10^7
μm^2/defect (30000 $mils^2$/defect).

Index